COMPUTATIONAL ORIENTED MATROIDS

Oriented matroids play the role of matrices in discrete geometry, when metrical properties, such as angles or distances, are neither required nor available. Thus they are of great use in such areas as graph theory, combinatorial optimization, and convex geometry. The variety of applications corresponds to the variety of ways they can be defined. Each of these definitions corresponds to a differing data structure for an oriented matroid, and handling them requires computational support, best realized through a functional language. Haskell is used here, and, for the benefit of readers, the book includes a primer on it. The combination of concrete applications and computation, the profusion of illustrations, many in colour, and the large number of examples and exercises will make this an ideal text for introductory courses on the subject. It will also be valuable for self-study for mathematicians and computer scientists working in discrete and computational geometry.

JÜRGEN BOKOWSKI is a professor of mathematics at Darmstadt University of Technology.

Computational Oriented Matroids

Equivalence Classes of Matrices within a Natural Framework

JÜRGEN G. BOKOWSKI

Darmstadt University of Technology
Germany

CAMBRIDGE
UNIVERSITY PRESS

Shaftesbury Road, Cambridge CB2 8EA, United Kingdom

One Liberty Plaza, 20th Floor, New York, NY 10006, USA

477 Williamstown Road, Port Melbourne, VIC 3207, Australia

314–321, 3rd Floor, Plot 3, Splendor Forum, Jasola District Centre, New Delhi – 110025, India

103 Penang Road, #05–06/07, Visioncrest Commercial, Singapore 238467

Cambridge University Press is part of Cambridge University Press & Assessment,
a department of the University of Cambridge.

We share the University's mission to contribute to society through the pursuit of
education, learning and research at the highest international levels of excellence.

www.cambridge.org
Information on this title: www.cambridge.org/9780521849302

First published 2006

A catalogue record for this publication is available from the British Library

ISBN   978-0-521-84930-2   Hardback

To Barbara, Maria, Martha and Alma

Contents

Preface

The theory of oriented matroids is like a flourishing flower bed within the garden of discrete mathematics. Most of its plants are entwined with the many branches of other fields. It is difficult, if not impossible, to determine the multitude of independent roots. However, by examining the blossoms yielded, we can build a picture of this flower bed; reaping the fruits of the theory of oriented matroids and all its applications without requiring the spectator to be a "gardener." Thus, we hope that the reader can enjoy this developing picture and appreciate the beauty of its growth.

This book provides an introduction to oriented matroids for mathematicians, computer scientists, and engineers. It contains basic material for a course on polytopes, discrete geometry, linear programming, robotics, or any subject in which oriented matroids play a role. Software on the subject supports not only the process of learning for the student but makes the book very valuable for any specialist in the field.

Let us consider an example. Take the end points $1, 2$ of a line segment and the vertices $3, 4, 5$ of a triangle. Does the line segment intersect the triangle? This question is fundamental and decisive in computer graphics, robotics, and many other geometric problems. To answer it, many people calculate the point of intersection between the line given by the line segment and the plane given by the triangle. They decide later whether the point of intersection lies within the triangle and within the line segment. In practical applications it is typical to have many decisions of this type. Based on the theory of oriented matroids, we see the answer depends on just 5 signs, the orientations of the tetrahedra formed by ordered subsets of 4 points.

Take a (5×4)-matrix M defining in homogeneous coordinates the end points $1, 2$ of a line segment and the vertices $3, 4, 5$ of a triangle in Euclidean 3-space. The intersection property is invariant under rigid motions, and it can be expressed by "forgetting" the concept of a matrix which does change under rigid motions:

we use the *oriented matroid* of M. The intersection property depends on just five signs, the signs of the determinants of all (4×4)-submatrices of M. Up to reversing all signs simultaneously, there is precisely one solution which corresponds to the intersection case, showing that this property is invariant under rigid motions.

For mathematicians, computer scientists, and engineers matrices have become so useful in all their applications that they are reluctant to give up such a concept in order to generalize it to that of an oriented matroid. But there are many reasons to do so, especially when the matrices are used to describe *geometric objects* (such as polyhedra, convex polytopes, hyperplane arrangements, finite point sets, etc.), and the investigation concerns *combinatorial properties* of their geometric objects.

An *oriented matroid* is a generalized rigid motion invariant matrix function of high geometrical relevance like the five signs in the above example. More than three decades of exciting development in mathematics together with early contributions since 1926 have created a *theory of oriented matroids* the fundamentals of which deserve to be known by mathematicians, computer scientists, and engineers because of their applications and simplifications.

Oriented matroids even turn out to be invariant under topological transformations of the projective space that require a definition leaving the framework of Euclidean geometry. But, instead of leading to a drawback, this fact has a decisive advantage: the invariant of the projective space can always be described as a finite sequence of signs, and this in turn allows computations in terms of simple sign considerations.

For engineers working with geometric objects, for example in robotics, we recommend that their algorithms are modified to ones, where pure sign calculations can replace and simplify their computations with real numbers. Just as complex numbers have proven useful in studying real zeros of polynomials, oriented matroids play a key role in the context of geometrical problems with combinatorial properties. The generalization of those rigid motion invariant matrix functions allows an inductive generation of a complete (super-) set of geometric objects with given combinatorial properties. For example, this fact has led to the solution of a longstanding open problem in geometry concerning triangulated closed orientable two-dimensional manifolds and their flat embeddings in 3-space.

By providing independent motivations for the study of oriented matroids, throughout the mathematical history of geometry, optimization, pseudoline arrangements, molecule classification, theory of ordinary matroids, zonotopal tilings, etc., the book develops fundamentals in the theory of oriented matroids, and it presents results in the theory of oriented matroids which have not been covered in books thus far.

From the very beginning the reader is invited "to learn by playing" with various programs for oriented matroids and small-scope examples on his or her PC or on any UNIX computer in order to avoid the burden of changing the different models for oriented matroids by hand. There are many models of oriented matroids corresponding to rather different data structures. A substantial part of explanations concerns the transition between these data sets. We have used the de facto standard functional programming language Haskell 98 to explain those transitions as well as several other aspects. These functions can be evaluated. We have written a short Haskell primer that contains the restricted part of Haskell 98 that we need in this book. It is useful to have a look at this Haskell primer first when the reader hears about this functional language here for the first time.

In Chapters 1 and 2 we look at geometric matrix models in many different ways. This stance is very useful for a profound understanding of the theory of oriented matroids. In Chapter 3 we extend the combinatorial way of looking at matrices to that of oriented matroids in rank 3. We reach the arbitrary rank case in Chapter 4. Thus the reader has some motivation from previous chapters when we provide rather abstract axiom systems for oriented matroids. In Chapter 5 we study face lattices of oriented matroids, especially the convex hull concept within the theory of oriented matroids. A reverse concept forms the topic in Chapters 6 and 7. Here we start with say a face lattice that might be that of a polytope. We discuss how to find an oriented matroid the convex hull of which has the given face lattice. Chapter 8 is devoted to a central question in the theory of oriented matroids. Given an oriented matroid, can we find a matrix the oriented matroid of which is the given one? These theoretical contributions have a lot of applications in problem classes of computational synthetic geometry. We look at them in Chapter 9. Some additional applications of the theory of oriented matroids form the content of Chapter 10 and some intrinsic problems of the theory will be discussed in Chapter 11. Functional programming forms our faithful companion for carrying the burden of handling the data structure of oriented matroids. Moreover, it enables a concise and profound understanding of both transitions between different oriented matroid models and algorithmic concepts within our applications. Thanks go to several people who have commented on the manuscript at various stages. I am grateful to my son Boris who persuaded me that functional programming is a very useful tool in this context.

It was of much help to be allowed to use computer graphics, photos, models, and artwork of Benno Artmann, Andy Goldsworthy, Norman Hähn, Erich Hartmann, Karl Heinrich Hofmann, Michel Las Vergnas, and Carlo Séquin within this volume. I am grateful to Gabriela Hein for her skilled guidance during my production of all my pottery models. Above all I am very delighted that Cambridge University Press has printed all those figures in colour in which the coloured version expresses so much more than words can do.

Jürgen G. Bokowski, September 2005

1

Geometric matrix models I

A matrix will be our paradigm for explaining the concept of an oriented matroid. In this chapter the matrix

$$
M := \begin{pmatrix} 0 & -1 & 1 \\ 0 & 0 & 1 \\ -1 & -1 & -1 \\ 0 & -1 & -1 \\ 1 & 1 & 0 \end{pmatrix}.
$$

will be our companion when we ask: "what does a matrix represent?"

There is a surprising multitude of definitions for an oriented matroid. Starting with a certain collection of definitions does not at the beginning really help us to understand why we should study them. Moreover, some of these definitions are so different from others that we are faced with the problem of proving their equivalence. This is seldom an easy task. The novice is in general more confused than motivated if the axioms are given without further explanation. Oriented matroids have many different historical roots. The theory grew independently for many years from various places.

Nowadays all these results have merged into one theory of oriented matroids. Some contexts in which they came up include topics like

- line and pseudoline arrangements, see Levi, 1926;
- Grassmannians, see Gutierrez Novoa, 1965;
- oriented graphs, see Las Vergnas, 1974, 1975, 1978, 1981;
- linear programming, see Bland, 1974;
- topological sphere systems, see Folkman and Lawrence, 1978;
- semispaces and configurations, see Goodman and Pollack, 1984;
- convex polytopes, see Bokowski, 1993;
- molecule classification, see Dreiding, Dress, and Haegi, 1982;
- point configurations in chemistry, see Klin, Tratch, and Treskov, 1989;

- order functions, see Kalhoff, 1996 and Jaritz, 1996;
- Platonic Solids, see Bokowski, Roudneff, and Strempel, 1997;
- hull systems, see Knuth, 1992.

to mention but a few of them. A single abstract definition of an oriented matroid, as desirable as it might be, hides from the novice its many different aspects which, as a whole, sum up to a mighty tool in discrete geometry. Thus, we have chosen another approach which will hopefully lead to a more profound understanding and better introduction: we study matrices first. Matrices, actually certain equivalence classes of them, serve as good examples for oriented matroids.

When picking a definition to begin with, we have to start with a particular concept. We may immediately find a nice idea and intriguing applications but we thereby lose the overview. In the theory of oriented matroids there are many ways to switch from one concept or model to another. Each change offers new ideas and casts new light onto a problem allowing the reader to reformulate it. Thus, for our introduction, we recommend that the reader does without a precise definition. At this early stage, we do say that we are going to generalize equivalence classes of matrices. We will find a natural framework to work with these classes. For those aquainted with the (ordinary) matroid concept and wishing to understand the theory of *oriented* matroids, we provide a reason for the word *natural*. It expresses that we do not include all matroids. Matroids have been studied extensively for other reasons before the theory of oriented matroids became an important theory in its own right. Many matroids, that is, the non-orientable ones, are canceled out in the theory of oriented matroids, thereby gaining, for example, the convexity properties of oriented matroids.

Once again, looking in more detail at familiar matrix concepts and geometrical models is our crucial starting point for a better understanding of the fundamentals in the theory of oriented matroids.

What is an $(n \times r)$-matrix of maximal rank r with real coefficients? What does it describe, what does it stand for, what does it represent?

- Is it the vector space V generated by the column vectors of the matrix?
- Does it describe the vector space in $\mathbb{R}^n$ which is orthogonal to V?
- Is it a set of vectors given by the n rows of the matrix?
- Does it describe a zonotope given by the rows of the matrix?
- Does it describe the projection of an n-cube?
- Is it a set of normal vectors of a central hyperplane arrangement?
- Does it describe a point set of homogeneous coordinates?
- Does it describe a linear map?
- Are the rows of the matrix the vertices of a convex polytope?
- Does it represent a point on a Grassmannian?
- Does it tell us all Radon partitions of a point set?

This list of questions is not complete. Nevertheless, it indicates why a definition for an oriented matroid can appear in so many forms. Linear algebra is a basic subject in many areas of mathematics and its applications. The concept of a matrix is fundamental in these contexts. A matrix leads us to a paradigm of an oriented matroid.

The oriented matroid of a matrix can be viewed as an invariant. Like the edge graph of a convex polytope, an oriented matroid is invariant under rigid motions. We can also compare an oriented matroid with the f-vector of a convex polytope. The f-vector, $f = (f_0, f_1, \ldots, f_i, \ldots)$, with components $f_0 =$ number of vertices, $f_1 =$ number of edges and in general $f_i =$ number of i-faces, is also an invariant under rigid motions of the polytope. These concepts do not even change if the convex polytope is replaced with a homeomorphic image of it, a topological ball, having the same boundary structure.

If the matrix is thought to describe a set of outer normal vectors of hyperplanes bounding a convex polytope, it also defines an invariant of this polytope, a topological ball that can also be obtained if the polytopal cell is bounded by topological hyperplanes. This way of thinking comes very close to the concept of an oriented matroid. However, the definition of an oriented matroid can be given in many contexts corresponding to the above questions and answers.

There is resistance to generalizing the concept of a matrix. If we think of the complex numbers, which have proven useful for studying zeros of polynomials, we can understand that a similar completeness property can require a more general concept. In the case of matrices this will be a topological invariant. We arrive at an invariant within projective space that cannot be represented by a matrix in all cases.

By introducing such a general concept, we gain something important. We can generate within this generalized framework all combinatorial types of abstract point configurations, with a given number of points, for a given dimension and sometimes even with certain prescribed properties. Later we can sort out those configurations that are of interest to us. We obtain with combinatorial methods an overview of all point configurations. This can be seen in many instances as the paramount advantage. Often there are no other methods available to get a similar overview.

After these general remarks, we look at matrices and what they represent.

In the first two chapters we have listed various ways to present partial or full information of a given matrix. Our catalogue is not complete but it shows some contexts in which oriented matroids appear.

Section 1.1 about convex polytopes was chosen first because oriented matroids form an essential tool for investigations in the theory of convex polytopes. The concept of polar duality as a well-known example of an anti-isomorphism of the face lattice of a convex polytope with its polar dual polytope serves us as an example. Similar isomorphisms occur in many forms in the theory of oriented matroids. Vector configurations in Section 1.2 play an essential role in problems in computational geometry. In Section 1.3 the chirotope will be the first pure data structure of the oriented matroid of a matrix. In Section 1.4 we see the polar dual aspect of Section 1.2. It will turn out later that the polarity concept cannot be carried over to the theory of oriented matroids, in general. Nevertheless, applications of oriented matroids play a role in both cases. In particular, we obtain the sphere arrangements of a matrix as an easy transition from the central hyperplane concept. Moreover, the equivalence class of homeomorphic transformations of such sphere systems leads to a second data structure of the pure oriented matroid information of a matrix. We introduce a third data structure for oriented matroids of a matrix in Section 1.6. It combines the more geometrical flavor of a sphere system with the computational advantage of a compact data structure like the chirotope. A set of hyperline sequences has proven very useful for finding extensions of a given oriented matroid with prescribed properties. Point sets on a sphere contain the matrix information up to a normalizing factor for each row vector. If we interpret these row vectors as unit normal vectors of central hyperplanes, their intersection with an affine hyperplane gives an ordered set of oriented hyperplanes. The duality concept, different from the polar dual concept, is a key tool in oriented matroid theory. We get a first understanding in Section 2.1 together with a tool that transforms the given chirotope into its dual one. We will see in our next chapter that chirotopes and sphere systems are just two different isomorphic forms in which an oriented matroid can be represented in the general case. There must not always be a corresponding matrix leading to both concepts. The pair of dual chirotopes, the pair of dual sphere systems, and the pair of dual hyperline sequences together with the pair of dual cocircuits are main building blocks for understanding the calculus of oriented matroids.

The concept of the Grassmannian in Section 2.2 is essential to an understanding of the chirotope axioms from a higher point of view. The sphere system induces a cell decomposition in the corresponding projective space. We discuss this aspect in Section 2.3. Closely related is the covector concept in Section 2.4. It describes the big face lattice of the corresponding cell decomposition. Oriented matroids are good objects to investigate zonotopes. This is the main message of Section 2.5 and in a slightly different form in Section 2.6. Cocircuits and circuits appear once more in a different setting in Sections 2.7 and 2.8. A summary section finishes our second chapter.

1.1 A convex polytope

In this section we introduce some easy concepts of convex polytopes. We try to indicate again why we do not begin immediately with a definition of an oriented matroid.

Consider the following key observation. Whereas a polytope is often given by a matrix, a data structure that does change, for example, under rigid motions, we are often interested in a property of the polytope, that is invariant under rigid motions, for example, its edge graph. If we calculate similar invariant information from the polytope, where do we lose the actual matrix information? Does the matrix lead us to a nice invariant?

In this book we will finally answer the last question in the affirmative: the oriented matroid of the matrix will contain the essential invariant information that we are interested in. However, before we understand this from a higher point of view, let us start with some aspects concerning polytopes.

We are going to investigate the transition from a polytope to its polar dual. This exemplifies how combinatorial information can have at least two representations.

$$M := \begin{pmatrix} 0 & -1 & 1 \\ 0 & 0 & 1 \\ -1 & -1 & -1 \\ 0 & -1 & -1 \\ 1 & 1 & 0 \end{pmatrix}.$$

Later in this chapter, we consider matrices on a more general level. We study not only a pair of representations of a matrix but a multitude of them. The oriented matroid will be the representation of the matrix that serves for all versions as the actual invariant form. However, the resulting data structure depends very much on the context in which the matrix was used.

We define a convex polytope P in Euclidean space $\mathbb{R}^d$ as the convex hull of a finite point set $A = \{x_1, \ldots, x_n\}$. We assume that our polytope is full dimensional, that is, it does not lie in a hyperplane.

$$P := \text{conv}\{x_1, \ldots, x_n\} := \{x \mid x = \sum_{i=1}^{n} \lambda_i x_i, \sum_{i=1}^{n} \lambda_i = 1, \lambda_i \geq 0\}.$$

If we have a hyperplane $H = \{x \mid \langle v, x \rangle = a\}$ in $\mathbb{R}^d$ with unit normal vector $||v|| = 1$, such that $F := H \cap P \neq \emptyset$ and $P \subset H^- := \{x \mid \langle v, x \rangle \leq a\}$, that is, the hyperplane touches the polytope and the closed halfspace in the opposite direction to v contains the whole polytope, we call H a *supporting hyperplane* of P and we call F a *face* of P. The *affine hull* of F aff $F := \{x \mid \sum_{i=1}^{d} \lambda_i y_i, \sum_{i=1}^{d} \lambda_i = 1, y_i \in F\}$ has as dimension, dim aff F, the dimension of the corresponding linear space parallel to aff F. This dimension is also used as the dimension of the face F. We have in particular the *vertices* of P as the zero-dimensional faces, the *edges* as

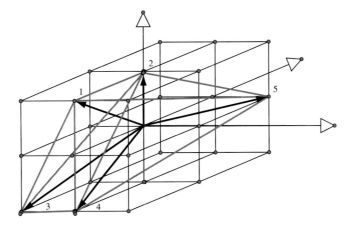

Figure 1.1   Matrix M representing a polytope

the one-dimensional faces of P, and the *facets* as the $(d-1)$-dimensional faces of P.

We reduce the set A to the *vertex set* vert $P := \{y | y \text{ is a vertex of } P\}$ of P. We assume that all points in A are vertices of P. We write the coordinates $x_i^1, \ldots, x_i^d$ of vertex x_i of the polytope P as the ith row of a matrix M.

$$
M_{\text{aff}}(P) := \begin{array}{c} 1 \\ \\ i \\ \\ n \end{array} \begin{pmatrix} x_1^1 & x_1^2 & \cdots & x_1^d \\ & \vdots & & \vdots \\ x_i^1 & x_i^2 & \cdots & x_i^d \\ & \vdots & & \vdots \\ x_n^1 & x_n^2 & \cdots & x_n^d \end{pmatrix},
$$

$$
M_{\text{hom}}(P) := \begin{array}{c} 1 \\ \\ i \\ \\ n \end{array} \begin{pmatrix} 1 & x_1^1 & x_1^2 & \cdots & x_1^d \\ 1 & \vdots & & & \vdots \\ 1 & x_i^1 & x_i^2 & \cdots & x_i^d \\ 1 & \vdots & & & \vdots \\ 1 & x_n^1 & x_n^2 & \cdots & x_n^d \end{pmatrix}.
$$

More often we add an additional column vector with entries 1 in the matrix interpreting the rows as homogeneous coordinates of the points. The latter matrix of a polytope will be the one that we study in detail later.

Let us now look at the polarity concept. For presenting a polytope, a more unusual way is that of providing its supporting planes.

A matrix can describe both a set of vertices of a polytope and a set of supporting planes.

If a convex polytope P in Euclidean space is given as its set of vertices in matrix form, then it is called *V-presented*, and if it is given as its set of supporting hyperplanes as a matrix, then it is called *H-presented*.

If we compare in the triangle example of Figure 1.2, the parameters of the vertices (two coordinates in each case) with the parameters of the lines (two components of the outer normal vector and the oriented distance from the origin), the numbers of parameters seem not to match, but the length of the outer normal vector of the supporting lines does not count. The better model appears if we use homogeneous coordinates, that is, if we consider convex cones generated by convex polytopes.

We use another example of a convex polytope, just a line segment S, and we consider a cone C spanned by this line segment S. We can describe the cone C by either the two vectors on the right in Figure 1.3, by the extreme rays of this cone C, or by the two rays on the left in Figure 1.3. These are extreme rays of the polar dual cone C^*, which can be interpreted as outer normal vectors of the facets of the first cone C.

It is sometimes useful to look at the matrix of a *V-presented* polytope as being the matrix of a *H-presented* polytope, or vice versa, to study properties on this *polar dual* polytope and to reinterpret results as such of the original object.

In dimension 2 this is easy to imagine and we simply learn as in the case of a regular hexagon: vertices and edges of the polar dual objects interchange.

Now we take a three-dimensional cube in a three-dimensional hyperplane H not passing through the origin in Euclidean 4-space. Assume that the center of the cube is the nearest point in H from the origin. Let H^* be the other parallel hyperplane to H with the same distance to the origin. The cube generates a four-dimensional convex cone C with apex at the origin. If we use the polar dual cone C^* of C

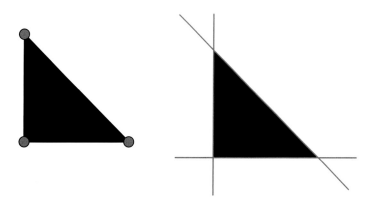

Figure 1.2 *V- and H-presented* triangle

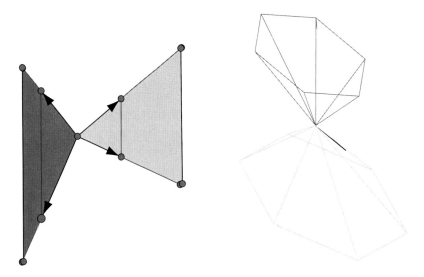

Figure 1.3   *V- and H-presented* cones, line segment, and regular hexagon

as shown in Figure 1.3 (but now in dimension 4) for which the extreme rays are formed by the outer normal vectors of facets of C, we obtain an octahedron as the intersection of the cone C^* with the hyperplane H^*. The octahedron is the polar dual convex body of the cube and vice versa, and this four-dimensional setting explains it best.

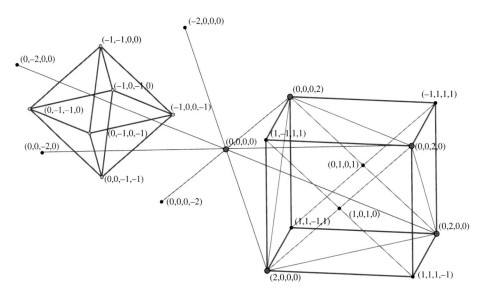

Figure 1.4   Polar duality, cube, and octahedron, projection from 4-space

The cube lies in the hyperplane $\sum x_i = 2$, the octahedron lies in the hyperplane $\sum x_i = -2$. We have used the formal definition of the polar body K^* of a non-empty, compact, convex set K with the origin in its interior:

$$K^* = \{x \in \mathbb{R}^d \mid \langle x, y \rangle \le 1 \text{ for all } y \in K\}.$$

Like the aforementioned polarity concept, there is a whole array of such transitions from various models of a matrix to others, especially if combinatorial properties of the matrix are involved. We are going to list matrix models in this chapter. Understanding these will help to reduce possible confusion for the novice when faced later with generalized concepts of matrices derived from different models.

If we study properties of convex polytopes, very often it is actually the rigid motion invariant information of that matrix, we are interested in. With the volume of the polytope, its edge graph or 1-skeleton, or more generally, with the whole face lattice of the polytope, we obtain a rigid motion invariant. The affine hulls of the facets of the polytope define a cell decomposition of the corresponding space. This cell decomposition and its face lattice (it is called the big face lattice of the polytope) is another invariant given by the polytope. If we study the cell decomposition given by the polytope, the original matrix also contains that information of the cell decomposition. It turns out that there are closely related cell decompositions that should be studied within the same framework and that cannot be represented by matrices. In that sense the matrix concept lacks a completeness property and thus has to be generalized. The idea of focussing on invariants under transformation groups, proposed by Felix Klein in his *Erlanger Programm*, guides us. We study in the following an invariant of matrices which does not even change under homeomorphic (i.e., bijective and continuous) transformations of the underlying projective space.

First, we discuss some geometric models for a matrix, or rather for equivalence classes of matrices. What are geometric or other models of a generic real $(n \times r)$-matrix? Generic here means: M has rank r and any pair of row vectors is linearly independent. We exemplify our answers for $n = 5$, $r = 3$ in the case of the following matrix M.

$$M := \begin{array}{c} 1 \\ 2 \\ 3 \\ 4 \\ 5 \end{array} \begin{pmatrix} 0 & -1 & 1 \\ 0 & 0 & 1 \\ -1 & -1 & -1 \\ 0 & -1 & -1 \\ 1 & 1 & 0 \end{pmatrix}.$$

The term E is used for the ground set of elements $E = \{1, \ldots, n\}$. Although we have already mentioned that for a polytope P we prefer to study a matrix with homogeneous coordinates $M_{\mathrm{hom}}(P)$, we can interpret the above matrix as representing a polytope with five vertices in ordinary space.

We conclude this section with a short summary. A polytope, given by a matrix, does change under rigid motions. As in the case of the edge graph of the polytope, we prefer an invariant representation. If we calculate rigid motion invariant information of the polytope, where do we lose the matrix information? A question of this type will lead us to the oriented matroid of the matrix. But for our polytope with 5 vertices the corresponding matrix has to be that where we use homogeneous coordinates:

$$M' := \begin{array}{c} 1 \\ 2 \\ 3 \\ 4 \\ 5 \end{array} \begin{pmatrix} 1 & 0 & -1 & 1 \\ 1 & 0 & 0 & 1 \\ 1 & -1 & -1 & -1 \\ 1 & 0 & -1 & -1 \\ 1 & 1 & 1 & 0 \end{pmatrix}.$$

Our matrix example with 3 columns implies only lower dimensional concepts. But, of course, we have applications in mind in which even the polytope dimension exceeds the familiar number three. The interested reader can try to work in all the next sections with the above polytope matrix M'.

1.2 A vector configuration

$$M := \begin{pmatrix} 0 & -1 & 1 \\ 0 & 0 & 1 \\ -1 & -1 & -1 \\ 0 & -1 & -1 \\ 1 & 1 & 0 \end{pmatrix}.$$

We continue with perhaps the most natural *vector model* of a matrix. Interpreting the $n = 5$ rows of the matrix as vectors in $\mathbb{R}^r$, or in our example in $\mathbb{R}^3$, we obtain an ordered set of vectors v_i, $i \in E$, in $\mathbb{R}^3$. For any choice of $r = 3$ row vectors (v_j, v_k, v_l) of the matrix M, $1 \le j < k < l \le n$, with increasing indices, we can check whether they form a basis of $\mathbb{R}^3$, that is, whether the determinant

$$[j, k, l] := \det \begin{pmatrix} v_j \\ v_k \\ v_l \end{pmatrix}$$ of the submatrix of M formed by the three rows j, k, l,

is non-zero.

In the affirmative case, we call the ordered tuple (j, k, l) a basis and by calculating the sign of the determinant $[j, k, l]$ of the submatrix, we obtain the orientation

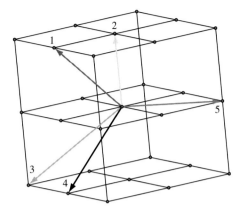

Figure 1.5 Vector configuration of the matrix M, see also Figure 1.17

of that basis. In our example we list the bases and their corresponding orientations together with the non-bases, that is, those tuples of indices that form no basis.

```
(1,2,3)basis sign[1,2,3] = +1      (1,4,5)basis sign[1,4,5] = +1
(1,2,4)      sign[1,2,4] =  0      (2,3,4)basis sign[2,3,4] = +1
(1,2,5)basis sign[1,2,5] = -1      (2,3,5)      sign[2,3,5] =  0
(1,3,4)basis sign[1,3,4] = +1      (2,4,5)basis sign[2,4,5] = +1
(1,3,5)basis sign[1,3,5] = +1      (3,4,5)basis sign[3,4,5] = -1
```

We mention the Steinitz–McLane exchange lemma for bases from linear algebra in this context. It says, if we delete in a basis $b = (j, k, l)$ one of its elements, say j, and if we have another basis $b' = (j', k', l')$, we can find an element from the basis b' which together with k and l again forms a basis. In our example, we can confirm the theorem on an abstract level just by using all the above tuples.

If we are given an abstract set of r-tuples of E (a potential set of bases $\mathcal{B}$) and if we can confirm the Steinitz–MacLane exchange lemma in this abstract way, we call the pair $(E, \mathcal{B})$ a *matroid of rank r*.

The equation

$$\mathcal{B} = \{(1, 2, 3), (1, 2, 5), (1, 3, 4), (1, 3, 5), (1, 4, 5), (2, 3, 4), (2, 4, 5), (3, 4, 5)\}$$

forms in our example a *matroid* $(E, \mathcal{B})(M)$ of the given matrix M of rank 3. Just a single check: delete 3 in the basis (1,3,5) and consider, for example, the basis (2,4,5). We find an element in the second basis, the element 2, such that (1,2,5) lies again in the set $\mathcal{B}$ of bases. Note that a matroid does not carry any information about orientations of bases. We have not defined oriented matroids so far. For the theory of (ordinary) matroids we refer the reader to comparitively old books on (ordinary) matroid theory, see, for example, Welsh, 1976. For the part that *independent sets* and (ordinary) matroids play in optimization, see, for example, Bachem, 1983.

Functional programming

The novice in the theory of oriented matroids should think at an early stage about computational applications. We do not expect the reader to become a programmer, but for particular applications, computers help a lot in the theory of oriented matroids.

The data structure of oriented matroids requires computational help unless we restrict ourselves to very simple examples. There are various ways to store the information of an oriented matroid. A common method is to use lists of signed vectors. Handling them can be boring if we do not have any computational aid.

If we investigate oriented matroid examples that occur in the literature, it is very likely that there was a computational search process involved for finding these examples. A better understanding of the subject comes from performing some computations. The question arises in what language to describe them. There are of course old programs available and most researchers in the theory of oriented matroids have done some programming in one or other programming language. We have decided for this book to use the functional programming language Haskell 98, named after the mathematical logician Haskell B. Curry.

We have added a primer about Haskell 98 for those who are not familiar with this language. Many mathematical faculties and computer scientist faculties nowadays teach Haskell 98 in the first term. We use only a small portion of the language. Writing your own program might be more difficult at the beginning, however, understanding our Haskell programs should be easier.

Haskell 98 is the result of a joint effort of people working in the functional programming community.

Our Haskell programs are usually not very fast compared with specialized programs written in, for example, C. However, our programs are easier to understand and they serve as a mathematical tool to understand the theory of oriented matroids.

By using a functional programming language, we have the advantage that we stay close to mathematical definitions and we can check the program easily. For example, the following Haskell function `isM` (is matroid?) does all the checkings for a potential matroid structure for us.

Potential bases of a matroid, is it a matroid?

```
pb::[[Int]]
pb=[[1,2,3],[1,2,5],[1,3,4],[1,3,5],
    [1,4,5],[2,3,4],[2,4,5],[3,4,5]]

isM::[[Int]]->Bool
isM b
=and[or[s(u(xs\\[x])[y])'elem'b|y<-ys]|xs<-b,ys<-b,x<-xs]
    where s = sort; u = union -- library functions
```

We read the above code as follows. The function `pb` is a constant function, a list of lists of elements, sorted triples of potential bases of a rank 3 matroid. This function `pb` provides the input for the second function `isM` which returns a boolean variable, `True` in case of a matroid and `False` otherwise. Thus the program for testing the matroid property is just a single line.

We take all tuples `xs` from the list of tuples `b`, and, independently, all tuples `ys` from the list `b`, and finally all elements `x` from `xs`. The functions `union` and `sort` are the usual union and sorting functions from the List module (List.hs) of Haskell. The function `and` returns `True` if we obtain `True` for all elements in the interior list of Boolean variables. The function `or` returns `True` if, in at least one case, the sorted tuple in the interior list has the `True` value, that is, the (sorted) union of `xs \\[x]` with `[y]` is an element in `b`.

Compare the matroid explanation with the executable program of just one line! Moreover, the function `isM` works in arbitrary dimensions. Of course, we have assumed that each potential base of the input data has been sorted.

What has to be done to execute the above program? Assume that your computer runs under Linux and the Hugs interpreter for Haskell 98 source code has been installed, otherwise see references given in the Haskell Primer. Write a file `MatroidCheck.hs` beginning with

How to get started

```
module MatroidCheck where
import List -- a library of list functions
```

followed by the two functions `pb` and `isM`. Now type `hugs Matroid-Check.hs` to get the hugs prompt and type `isM pb` to get the answer `TRUE`. In other words, the list of triples is indeed a matroid.

All *r*-tuples of a list of integers

```
tuples::Int->Int->[[Int]]
tuples 0 n = [[]]
tuples r n = tuplesL r [1..n]

tuplesL::Int->[Int]->[[Int]]
tuplesL r list@(x:xs)
   | length list < r = []
   | length list == r = [list]
   | r == 1      = [[el]|el<-list]
   | otherwise = [[x]++el| el<-tuplesL (r-1) xs]
                 ++tuplesL r xs
```

We very often deal in the following with tuples of integers. In the Haskell
primer, we have described such a function among the first recursive functions.
The function `tuples` returns all *r*-tuples of the list `[1..n]` of the first *n*
natural numbers. The function `tuplesL` does the same with any other set, that
is, we get all *r*-tuples from an arbitrary list. We can either use the variable
`list` or we use the head of that list as `x` or the tail of that list as `xs`. If the
length of `list` is equal to `r`, we return a list containing that list. If `r` equals 1,
we return a list of one element lists. Otherwise, we construct the list of tuples
recursively as those containing the head `x` and additional $(r-1)$ elements and
those that do not contain `x`. The sign "@" allows for using either "`list`" or
"`(x:xs)`".

The tuple function occurs very often because we consider *oriented simplices*
as pairs of a tuple and a sign. We allow the degenerate case where the simplex
has no maximal dimension and the sign is zero.

1.3   A chirotope, orientations of all simplices

The information about all $\binom{n}{r}$ oriented bases and all non-bases, sorted according
to the natural ordering of the *r*-tuples, can be stored as a sign vector, the so called
chirotope $\chi(M)$ of the matrix M. We obtain for our example matrix

$$M := \begin{pmatrix} 0 & -1 & 1 \\ 0 & 0 & 1 \\ -1 & -1 & -1 \\ 0 & -1 & -1 \\ 1 & 1 & 0 \end{pmatrix}$$

$$\chi(\mathbf{M}) := (+1, 0, -1, +1, +1, +1, +1, 0, +1, -1).$$

We should see this information in connection with `tuples 3 5`, that is, with the corresponding triples

$$[[1,2,3],[1,2,4],[1,2,5],[1,3,4],[1,3,5],$$
$$[1,4,5],[2,3,4],[2,3,5],[2,4,5],[3,4,5]].$$

The Haskell function `zip` from the List module applied to the lists

$$[1,0,-1,1,1,1,1,0,1,-1]$$

and the above tuple list `tuples 3 5` transforms this into the following.

Complete chirotope information of matrix *M*

```
zip (tuples 3 5) [1,0,-1,1,1,1,1,0,1,-1] =
          [([1,2,3], 1),
           ([1,2,4], 0),
           ([1,2,5],-1),
           ([1,3,4], 1),
           ([1,3,5], 1),
           ([1,4,5], 1),
           ([2,3,4], 1),
           ([2,3,5], 0),
           ([2,4,5], 1),
           ([3,4,5],-1)]
```

These lists of pairs will serve us as familiar data structure to look at oriented matroid information. The combinatorial information of the convex hull of the point set of all homogeneous row vectors of the matrix can be computed from this chirotope.

If this sign vector $\chi(M)$ is used, we speak of the *chirotope model*. The *oriented matroid given by the chirotope of the matrix* is the pair $(E, \chi)(M)$. Goodman and Pollack also introduced the term *order type* in this context.

A regular *n*-gon with counterclockwise labeled vertices turns into a sign vector with plus signs only. More generally, the chirotope of a cyclic polytope in any dimension can be represented as a sign vector with positive signs only, it is also called the alternating oriented matroid. We have used the function `altOM`. The type OB occurs very often, it is equal to `(Tu,Int)` or `([Int],Int)`, the pair of a tuple and a sign. For an overview about our types, see Page 303.

Alternating oriented matroid

```
altOM::Int->Int->[OB]
altOM r n=zip(tuples r n)(replicate(length(tuples r n))1)
```

The chirotope of M is an invariant with respect to the special orthogonal group $SO(3)$, the group of special rotations of the space $\mathbb{R}^3$ around its origin described via an orthogonal (3×3)-matrix R with determinant det $R = 1$. Multiplying M with an element $R \in SO(3)$ does not change the values of the determinant. The product of $[j, k, l]$ with det R is equal to $[j, k, l]$. In other words, if we multiply M with the inverse matrix of the submatrix of the first three row vectors, the resulting matrix has the same chirotope $\chi(M) = \chi(MR)$.

$$MR := \begin{pmatrix} 0 & -1 & 1 \\ 0 & 0 & 1 \\ -1 & -1 & -1 \\ 0 & -1 & -1 \\ 1 & 1 & 0 \end{pmatrix} \begin{pmatrix} +1 & -2 & -1 \\ -1 & 1 & 0 \\ 0 & 1 & 0 \end{pmatrix} = \begin{pmatrix} 1 & 0 & 0 \\ 0 & 1 & 0 \\ 0 & 0 & 1 \\ 1 & -2 & 0 \\ 0 & -1 & -1 \end{pmatrix}.$$

Next we observe that the chirotope information can be obtained from all signs of determinants of *all* submatrices (not only $(r \times r)$-*submatrices*) of the so-called *Tucker matrix T* of MR or the *standard representative matrix* of MR that we obtain in our example by deleting the unit matrix

$$T := \begin{pmatrix} t_{1,1} & t_{1,2} & t_{1,3} \\ t_{2,1} & t_{2,2} & t_{2,3} \end{pmatrix} = \begin{pmatrix} 1 & -2 & 0 \\ 0 & -1 & -1 \end{pmatrix}.$$

We have $\sum_{k=0}^{r} \binom{n-r}{k} = \binom{n}{r}$ such submatrices. The case $k = 0$ corresponds to the empty set or better to the above unit matrix.

Apart from the case, $\text{sign}(1, 2, 3) = \text{sign det} \begin{pmatrix} 1 & 0 & 0 \\ 0 & 1 & 0 \\ 0 & 0 & 1 \end{pmatrix} = 1$, we have

$\text{sign}(1, 2, 4) = \text{sign } t_{1,3} = 0, \ldots, \text{sign } (3, 4, 5) = \text{sign det} \begin{pmatrix} 1 & -2 \\ 0 & -1 \end{pmatrix} = -1$. We obtain the full chirotope information of M from the Tucker matrix $T(MR)$. Once more: for n points in Euclidean space $\mathbb{R}^{d-1}$: $x_i = (x_{i,1}, x_{i,2}, \ldots, x_{i,d-1})^t \in \mathbb{R}^{d-1}$, $1 \leq i \leq n$ and their matrix M of homogeneous coordinates, the *chirotope of a matrix M*, $\chi(M)$, is defined componentwise as an image of a d-tuple, as the sign of the determinant of the corresponding submatrix defined by that d-tuple. When we do not emphasize the particular data structure of the chirotope (of a matrix) we can also speak of the oriented matroid (of a matrix).

For points in general position the determinants are all non-zero. For this case (non-zero entries in a sign vector), we can say when such a data structure forms a general chirotope by definition.

For

$$\Lambda(n, r) := \{(\lambda_1, \ldots, \lambda_r) \in E^r \mid 1 \leq \lambda_1 < \cdots < \lambda_r \leq n\}$$

let $\chi : \Lambda(n, r) \to \{-1, +1\}$ be a map, extended onto all r-tuples out of n elements according to the determinant rules (alternating and anti-symmetric). We require that each restriction of χ onto $r+2$ of the n elements can be obtained as a chirotope of a matrix with $r+2$ rows (we do not require to have submatrices of a matrix), then we call the map χ a *"uniform" chirotope* or a *"uniform" oriented matroid* with n elements in rank d.

1.4 A central hyperplane configuration

What are other models for a generic $(n \times r)$-matrix? Interpreting the n rows of the matrix as normal vectors of central hyperplanes, leads to an ordered set of n central oriented hyperplanes E_i, $i \in E$, in $\mathbb{R}^r$.

If we choose r such oriented hyperplanes, or $r = 3$ such oriented planes with labels (j, k, l) in our example, we can look at the cone C, $C = C_{j,k,l}$ in our example, defined as the intersection of its corresponding positive halfspaces. If this cone C, or $C = C_{j,k,l}$, intersected with a unit sphere S^{r-1}, or $S^{r-1} = S^2$, around the origin, defines a non-degenerated spherical simplex, or triangle, the corresponding normal vectors form a basis of $\mathbb{R}^r$, or $\mathbb{R}^3$.

$$M := \begin{array}{c} 1 \\ 2 \\ 3 \\ 4 \\ 5 \end{array} \begin{pmatrix} 0 & -1 & 1 \\ 0 & 0 & 1 \\ -1 & -1 & -1 \\ 0 & -1 & -1 \\ 1 & 1 & 0 \end{pmatrix}.$$

There are two possible orientations of such a non-degenerated spherical simplex, or triangle. In our example, if looking from the outside, the planes with indices (j, k, l) appear either in clockwise or counter-clockwise order.

Let us consider the fixed set of n ordered oriented central hyperplanes. For each non-zero vector v and each oriented central hyperplane H_i, we can determine whether v lies on H_i, on the positive side of H_i, or on the negative side of H_i. We write the result as a corresponding signed vector for v. The example $(+, 0, -, +, 0)$ would tell us that there are five oriented hyperplanes and v lies on H_2 and on H_5, on the positive side with respect to H_1 and H_4 and on the negative side with respect to H_3. Can you find the corresponding cone in which v has to lie?

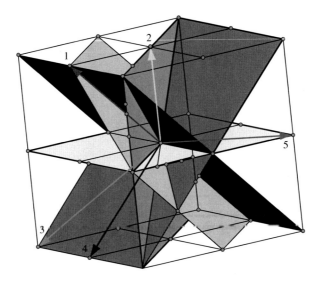

Figure 1.6   A central hyperplane configuration given by M, see also Figure 1.17

If we vary v, we obtain all possible signed vectors for the given matrix. The set of all possible signed vectors, if v varies, provides (among many other possibilities) a data structure that carries the oriented matroid information of the given matrix. The attempt to characterize those sets of signed vectors that can be obtained from all matrices leads us to generalize the matrix concept. However, such a formal definition, based on our signed vectors, will be given later.

1.5   A great-sphere arrangement

$$M := \begin{pmatrix} 0 & -1 & 1 \\ 0 & 0 & 1 \\ -1 & -1 & -1 \\ 0 & -1 & -1 \\ 1 & 1 & 0 \end{pmatrix}.$$

Interpreting the rows of our matrix M as before as normal vectors of central hyperplanes, or planes in the example, and cutting the unit sphere S^{r-1} around the center with these hyperplanes, or planes, we arrive at an ordered set of oriented great $(r-2)$-spheres, or oriented great circles, c_i, $i \in E = \{1, \ldots, n\}$ on the unit $(r-1)$-sphere.

The positive half on the sphere is that one to which the vector points. We call this model the linear *sphere system* model.

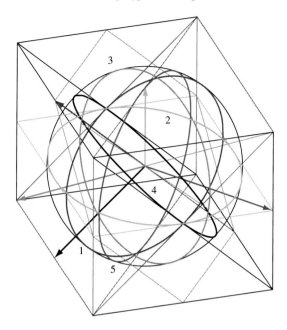

Figure 1.7   Circles on a sphere given by M, see also Figure 1.18

1.6   A set of hyperline sequences

In our example we walk along the oriented circles looking from the outside and having the positive side always to our right. If we have to cross other oriented great circles, we collect at the crossing point all signed indices of those circles crossing at the same time. We write the positive sign before the index if we change from the corresponding negative halfspace to its positive halfspace and the negative sign in the other case. In the small example of Figure 1.8, we would get locally

$$([1], [[+2], [-3]])$$

For our example matrix M, we obtain the following.

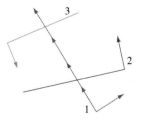

Figure 1.8   Small example

Set of circular sequences for our sphere system.

```
[( [1], [ [+2,-4], [-5], [+3], [-2,+4], [+5], [-3] ]),
 ( [2], [ [+1,+4], [-3,+5], [-1,-4], [+3,-5]        ]),
 ( [3], [ [+1], [+2,-5], [-4], [-1], [-2,+5], [+4] ]),
 ( [4], [ [+1,+2], [-5], [+3], [-1,-2], [+5], [-3] ]),
 ( [5], [ [+1], [+4], [-2,+3], [-1], [-4], [+2,-3] ])]
```

We have started in each circular sequence with the set containing the smallest positive element. We have used this as a normalization. Half of a sequence already determines the whole sequence. The other half of the sequence contains the opposite signs in the same order. Thus, we can skip the second half of the sequences and we arrive at the following.

Hyperline sequences

```
[( [1], [ [+2,-4], [-5], [+3] ]),
 ( [2], [ [+1,+4], [-3,+5] ]),
 ( [3], [ [+1], [+2,-5], [-4] ]),
 ( [4], [ [+1,+2], [-5], [+3] ]),
 ( [5], [ [+1], [+4], [-2,+3] ])]
```

given by the matrix

$$
M := \begin{matrix} 1 \\ 2 \\ 3 \\ 4 \\ 5 \end{matrix} \begin{pmatrix} 0 & -1 & 1 \\ 0 & 0 & 1 \\ -1 & -1 & -1 \\ 0 & -1 & -1 \\ 1 & 1 & 0 \end{pmatrix}.
$$

Again, let us look at all triples (i, j, k) of indices, $i, j, k \in \{1, \ldots, n\}$. The chirotope information $(+, 0, -, +, +, +, +, 0, +, -)$ from the vector model is also stored in this model of hyperline sequences. A pair of indices (j, k) occurs in the ith row in the same set if and only if the corresponding ordered triple $(a, b, c) = \pi(i, j, k)$, $a < b < c$, obtained by a permutation π from the indices i, j, k, is a non-basis. Now consider in the normalized ith row the set A_j, containing the index j with sign $s_j \in \{-1, +1\}$ and A_k containing the index k with sign $s_k \in \{-1, +1\}$. If A_j comes before A_k in the normalized sequence, we define $\text{sign}[i, s_j j, s_k k]$ to be $+1$, -1 otherwise. This leads to a well-defined sign of determinant function by requiring the function to be anti-symmetric, that is, we change signs if and only if the permutation π is not even. In other words, we can use $\text{sign}[i, s_j j, s_k k]$ to define $\text{sign}[a, b, c]$ consistently. What do we get for, for example, $(a, b, c) = (2, 3, 4)$?

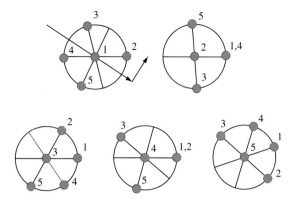

Figure 1.9 Hyperline sequences data structure via rotating oriented central lines

We look at the rows 2,3,4 and obtain sign$[2, +4, -3] := 1$, sign$[3, +2, -4] := 1$, sign$[4, +2, +3] := 1$. These signs are consistent and lead to sign$[2, 3, 4] = +1$. The reader can verify in our example that we get in all cases the well-defined sign of determinant function from our sequences and vice versa, the hyperline sequences can be obtained uniquely from the former list of bases and non-bases.

Normal representation of oriented simplex

```
norm::OB->OB
norm (tu,s) = normPos (list,s*signum prod)
 where prod = product tu; list = map abs tu
       normPos::OB->OB
       normPos (tuple@(h:rest),sign)
        |rest == [] = ([h],sign)
        |h==minimum tuple=([h]++fst next,snd next)
        |odd (length rest) =normPos (rest++[h],-sign)
        |otherwise         =normPos (rest++[h], sign)
        where next = normPos (rest,sign)
```

The function `norm` takes a signed tuple and returns it with its sorted tuple whereby the sign has been changed according to the alternating rule for determinants. When the signed tuple has an odd number of negative entries, we can use the absolute entries when we change the sign. Now the function `normPos` will be applied. For a 1-tuple there is nothing to do. When the tuple starts with the minimal element, we can apply the function `normPos` to the smaller tuple. Otherwise, we append the first element to the tail of the tuple and use the function normPos again. We have to take into account that the sign depends on the length of that tail.

Deletion of an element or of a set of elements in a chirotope

```
delSetChi::[Int]->[OB]->[OB]
delSetChi [] chi = chi
delSetChi (h:rest) chi = delElChi h (delSetChi rest chi)

delElChi::Int->[OB]->[OB]
delElChi el [] = []
delElChi el (h@(tu,_):rest)
  |el'elem'tu = delElChi el rest
  |otherwise  = [h]++delElChi el rest
```

The function delSetChi corresponds on the matrix level to deleting the rows in the matrix with those indices that are given as input. The function delElChi does the same for a single row index.

The function ctrSetChi can also be described on the level of a matrix. Assume for a moment that we have in the row with index i a unit vector having a 1 in the jth column. The effect of applying ctrSetChi [i] is to delete that row and delete the jth column. On an abstract level we use the words from point configurations: we consider a projection from that point i towards infinity. Applying the function ctrSetChi for a set simply means that we have projected again and again towards infinity thereby reducing the rank or the dimension each time by one. When there are dependences between the elements of the set, we have to discard them. This makes the corresponding program a little bit more complicated. We determine a maximal independent set of elements of the given set with the function findRankInd with respect to the given chirotope. At the same time we determine the rank.

In the case of a point configuration this means: we look at the affine hull of the given point set and we determine an affine base of it. The rank is the number of elements of this affine base. In the affine model language, we obtain with the function findRankInd a pair, the dimension together with an affine base of the affine hull. We continue with the language of point configurations that can be applied on this abstract level. The bases [fst st|st<-chi,snd st'elem'[-1,1]] are the simplices of maximal dimension of our chirotope. We do not consider degenerated simplices, that is tuples with orientation zero. The rank of the chirotope is given by length(fst (head chi)). When the given set occurs as a subset of a base of the chirotope, the result of findRankInd is clear. Otherwise, we can drop some element in `set` and we use the function `findRankInd` for the smaller set again. This has been written in the Haskell program.

Finding the rank and a base of a flat in a chirotope

```
findRankInd::[Int]->[OB]->(Int,[Int])
findRankInd set chi
  |or[length(base\\set)==r-k|base<-bases]= (k,set)
  |otherwise = maximum (map(i->findRankInd
             (take(i-1)set++drop i set)chi)[1..k])
  where k    =length set
        r    =length(fst (head chi))
        bases=[fst st|st<-chi,snd st'elem'[-1,1]]
```

For the contraction of a set we delete first all elements that are not in the determined base of the corresponding flat. Afterwards, we contract at all elements of the base.

The contraction of the chirotope at one element is formed by first deleting all tuples that do not contain that element. In addition we discard in all other tuples the given element and we reverse the orientation of the resulting tuple when the position of that element in the tuple was odd. Compare the function `ctrST` for this.

Contraction of an element in a chirotope

```
ctrElChi::Int->[OB]->[OB]; ctrElChi k [] = []
ctrElChi k (h:list) = ((ctrST k h)++ctrElChi k list)
  where ctrST::Int->OB->[OB]
        ctrST k (tu,sign)
          |k 'notElem' tu   = []
          |posk 'mod' 2 == 0 = [(tu\\[k],   sign)]
          |otherwise        = [(tu\\[k], -1*sign)]
          where posk = head (elemIndices k tu)
```

Contraction of a set of elements in a chirotope

```
ctrSetChi::[Int]->[OB]->[OB]
ctrSetChi [] chi = chi
ctrSetChi set@(h:rest) chi
  |k == length set = ctrSetChi rest (ctrElChi h chi)
  |otherwise = ctrSetChi list (delSetChi (set\\list) chi)
  where (k,list)=findRankInd set chi
```

Closure of an abstract flat

```
cl::[Int]->[OB]->[Int]
cl set chi = set++[e| e<-(els\\set), eInClSet e set chi]
  where els = sort(nub(concat(map fst chi)))


eInClSet::Int->[Int]->[OB]->Bool
eInClSet e set chi
  |rankE == rankSet = True
  |otherwise        = False
  where  rankSet = fst (findRankInd set chi)
         rankE  = fst (findRankInd ([e]++set) chi)
```

The function `cl` determines the closure of a set, that is, we find all points that lie in the affine hull of the given point set. This concept remains on the oriented matroid level. You can see an implementation from the corresponding Haskell program.

We provide a function that returns the matroid if the hyperline sequences are given. The function `hyperlines` is a constant function that just contains our example. The function `hyp2Matroid` returns the bases of our matroid. We have again used the function `sort` from the Haskell module `List`.

Hyperlines of our rank 3 example

```
hyperlines::[([Int],OM2)]
hyperlines = [ ( [1], [ [2,-4], [-5], [3] ] ),
               ( [2], [ [1, 4], [-3,5]    ] ),
               ( [3], [ [1], [2,-5], [-4] ] ),
               ( [4], [ [1, 2], [-5], [3] ] ),
               ( [5], [ [1], [4], [-2,3]  ] )]
```

Matroid information from hyperlines

```
hyp2Matroid::[([Int],OM2)]->[[Int]]
hyp2Matroid [] = []
hyp2Matroid (h:rest)
   = sort (nub ([ sort [abs a, abs b, abs c]
       |a <- fst h,
        i <- [1..((length(snd h))-1)],
        b <- ((snd h)!!(i-1)),
        j <- [(i+1)..(length(snd h))],
        c <- ((snd h)!!(j-1))]++hyp2Matroid rest))
```

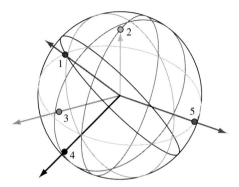

Figure 1.10 Points on a sphere given by M, see also Figure 1.18

The result of `hyp2Matroid hyperlines` is

$$[[1, 2, 3], [1, 2, 5], [1, 3, 4], [1, 3, 5], [1, 4, 5], [2, 3, 4], [2, 4, 5], [3, 4, 5]]$$

and `isM test` applied to this result returns `True`.

1.7 A point set on a sphere

$$M := \begin{pmatrix} 0 & -1 & 1 \\ 0 & 0 & 1 \\ -1 & -1 & -1 \\ 0 & -1 & -1 \\ 1 & 1 & 0 \end{pmatrix}.$$

The oriented great circles on the unit sphere are determined by corresponding points s_i, $i \in \{1, \ldots, n\}$ on the unit sphere, the intersection points of the unit sphere S^r, or $S^2 = \{x | x \in \mathbb{R}^3, ||x|| = 1\}$, with corresponding half lines generated by the vectors v_i, $s_i \in \{x | \ ||x|| = 1, x = \lambda v_i, \lambda \geq 0\}$. We refer to this model as *point configuration on the sphere*. Starting with an arbitrary matrix, by using the point system on the sphere, simply means that we have normalized all row vectors of the matrix. Multiplying the row vectors with a positive number does not change the information of the matrix in which we are interested.

1.8 A signed affine point set

$$M := \begin{pmatrix} 0 & -1 & 1 \\ 0 & 0 & 1 \\ -1 & -1 & -1 \\ 0 & -1 & -1 \\ 1 & 1 & 0 \end{pmatrix}.$$

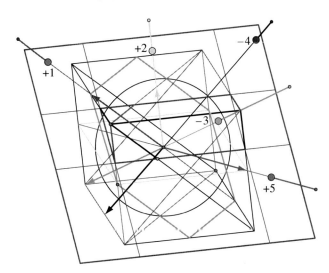

Figure 1.11   Signed points in a tangent plane of the ball, see also Figure 1.19

The two affine models we study in this and the next section are very often the starting point for investigations. Perhaps one is tempted to keep everything within this framework. However, embedding these affine problems in a linear framework can simplify the arguments. Later on, a final re-interpretation in the affine case might lead back to the original questions.

If we choose a tangent plane T of the unit sphere S^2, we can store the information of our ordered sets of vectors, oriented planes, oriented great circles, or

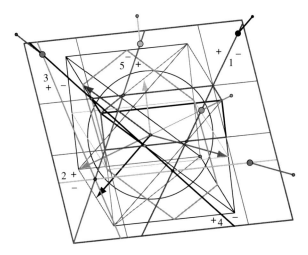

Figure 1.12   Oriented lines in a tangent plane of the ball, see also Figure 1.19

points on the unit sphere as signed points or oriented lines within that tangent plane. That is, as long as the vectors do not lie in a special position with respect to the chosen tangent plane T.

In other words, we choose for our given matrix a tangent plane T in general position, that is, no row vectors are parallel or orthogonal to the tangent plane. Then our given matrix can be interpreted as an ordered set of *signed points in an affine plane T* or as an ordered set of *oriented lines in an affine plane T*. In the first case, we consider points in T which lie in the oriented central lines defined by the vectors and we use a positive sign if the vector is a positive multiple of the vector, and a minus sign otherwise.

1.9   A set of oriented affine hyperplanes

We come back to our last remark in the previous section.

$$M := \begin{pmatrix} 0 & -1 & 1 \\ 0 & 0 & 1 \\ -1 & -1 & -1 \\ 0 & -1 & -1 \\ 1 & 1 & 0 \end{pmatrix}.$$

We consider intersections of oriented planes with the tangent plane T to obtain oriented lines in T. The correspondence of the points and the lines in the last two interpretations of matrix models is refered to as *polarity*. This concept has been mentioned in the context of *H- and V-presented* convex polytopes and polar duality, in Section 1.1. In Figure 1.13 we have depicted the five elements in both affine models without marking their orientations.

The following mathematical fairy tale underlines a decisive aspect of the theory of oriented matroids.

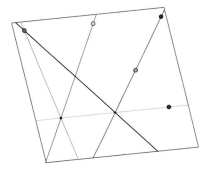

Figure 1.13   Lines and points of the given matrix M related by polarity, see also Figure 1.20

1.10 A magic globe

Once upon a time there lived five dwarves. Maybe there were more of them or fewer, no one can be sure in our time. Anyway, it is not important.

The dwarves were very shy. They feared nothing more than being watched in their cottages. Each one had his own little cottage to live in and an endlessly long straight path in the plane as his trail. All of the dwarves called the plane on which they lived a plateau since it lay high above a magic ball that they called their magic globe. For special events, the dwarves were allowed to come down to that magic globe and they always did this with great pleasure.

The magic ball had a special property. There was no way to tell apart a point on the ball from its opposite one. Therefore each pair of opposite points had just one name. A dwarf, standing on the globe, did so at the same time on the opposite point due to a magic doubling. Nonetheless, each dwarf counted only once.

Snow White watched everything with a beaming smile from the center of the ball. She kept in contact with the dwarves by shining rays of light into their cottages. They liked to be contacted in this way because each of them knew that was the sign that he was allowed to dance on the magic globe. He did so on the largest circle around the beam that pointed to his cottage. After a while his dancing circle on the magic ball could be seen by all. The magic property had the effect that after a half turn along the circle, the dwarf was back at his starting point.

It was easy to find the cottage of a dwarf from the dancing circle or from the straight path on the plateau. The reason for this was Snow White's beaming smile. When she looked to the plateau and followed a dancing dwarf on his circle, she saw a trail behind the dwarf. When she followed a dwarf on his trail, a line

Figure 1.14   Drawing: Karl Heinrich Hofmann, Darmstadt University of Technology

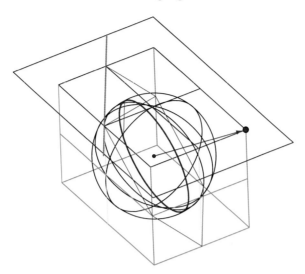

Figure 1.15 Plateau (blue) with cottage point (red), dancing circle (red), and trail (brown). Trail and circle lie in the same plane. A *Cinderella* drawing

orthogonal to all of the rays from Snow White to the trail would lead to the dwarf's cottage. Thus the trails and the cottages were connected in a nice way. When a dwarf seemed to vanish from his trail in one direction, he reappeared in the plateau from the opposite side.

All dwarves liked to meet and share the latest news with each other. Every pair of dwarves had a unique meeting point where their dancing circles met. Eventually, the dwarves memorized the sequence in which they met when they went in the usual direction along their dancing circles. They considered these sequences to be very good and they decided to keep them forever.

But alas, suddenly they realized that everyone could find their cottages by just watching their dances or their walks along the trails. What was to be done? They decided to deform the dancing circles but keep the sequences in which they met. If you are lucky enough to come to this magic world, try to find flat projected halves of this magic globe. You might find tracks of dancing dwarves. If they have not been erased, the dwarves run forever, happily ever after.

In Figure 1.16 we find projected half globes with visible traces. You can check the intersection property. Any two traces meet precisely once where they cross. For specialists in the field we provide hints about the origin of these pictures. From left to right and from top to bottom we have the one-element contraction of a dodecahedron, Harborth's example in which the number of triangles is maximal, a uniform representation of the bad pentagon example of Goodman and Pollak, and finally an example that was found during an investigation of the author with Roudneff (unpublished) in which all triangles are pairwise disjoint.

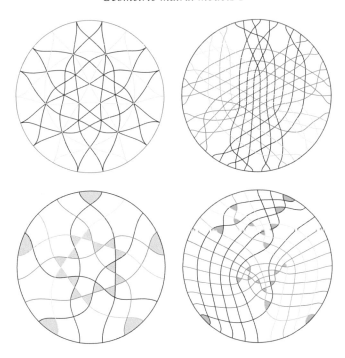

Figure 1.16 Projected half globes with visible traces

See Section 2.3 for remarks about the corresponding software *omawin* that was an outcome of a German Research Council project of the author's from the late eighties. It has been freely available for many years from `http://juergen.bokowski.de`. Essential parts of this software were written by Jürgen Richter-Gebert, Peter Schuchert, Peter Engels, and further research students in close contact with the author.

First summary

Figure 1.17 illustrates a polar dual pair of a geometrical interpretation of our sample matrix, vectors or central planes. Figure 1.18 shows again such a polar dual pair. The corresponding unit vectors define points on the 2-sphere and the central planes define great circles on the 2-sphere.

The transition from a chosen model to another one is not difficult in each particular case but the current model has to be specified. A lot of confusion can arise if the model of reference is not clear from the very beginning. This cannot be over-emphasized for the novice who wants to learn the fundamentals of the theory of oriented matroids. Do we consider the rows of a matrix to represent an $(r-1)$-polytope by its set of vertices in $\mathbb{R}^{r-1}$ or do we think of these rows as vertices given by homogeneous coordinates in $\mathbb{R}^r$? In other words, do we consider the point model in $\mathbb{R}^{r-1}$ or in $\mathbb{R}^r$? Do we consider the row vectors of a matrix

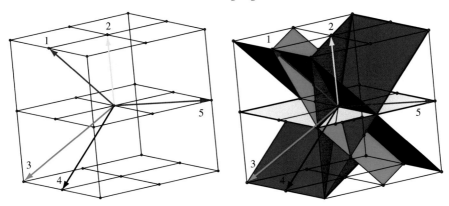

Figure 1.17   Vectors and central planes in $\mathbb{R}^3$

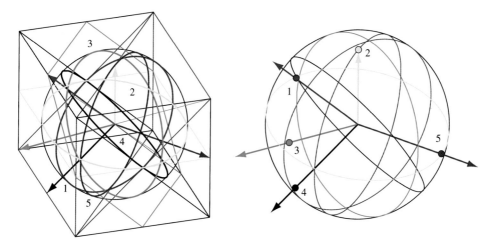

Figure 1.18   Sphere system and points on the 2-sphere

to represent an $(r-1)$-polytope determined by its set of supporting hyperplanes where the rows function as outer normal vectors of these hyperplanes? Do we consider the oriented hyperplane model which leads to a convex cone C in $\mathbb{R}^r$?

There are more such models still to come. For a better understanding of various equivalent systems of axioms for oriented matroids, it is convenient in each case to look at those geometric models which are close to the underlying abstract concept. Although the mathematical definition is clear in each case of a particular system of axioms, a creative working activity can be supported very much by example classes which have the right intuitive background.

Contractions of cubes and other polytopes

Before we start with a second series of representations for our given matrix, we work on another example that comes from studying a cube problem. Our vectors

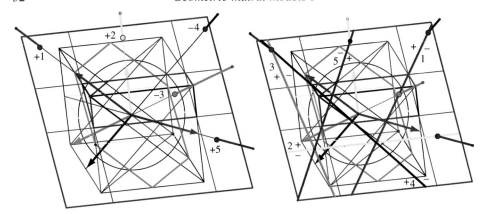

Figure 1.19   Signed points and oriented lines

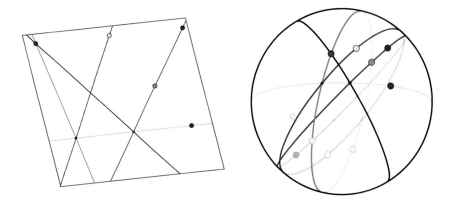

Figure 1.20   Affine and spherical picture, unoriented, unlabeled

are those that start at a fixed vertex of a cube and point to all remaining vertices, see Figure 1.21, on the left. Consider the cube such that Vector 1 stays vertical. The corresponding planes and great circles have been depicted in Figure 1.22. The elements 2, 4, 7 in Figure 1.21 form the upper "triangle" in Figure 1.22.

Finally we have shown the corresponding boundary structure on the upper half sphere projected onto the plane in Figure 1.21, on the right. Element 1 becomes the equator and we see the triangle 2,4,7 in the middle. This additional representation of our matrix is among the models still to come. This cube example was the starting point for an investigation of Las Vergnas's cube problem in higher dimensions, see Section 11.3.

We study the four-dimensional cube in the same way as the 3-cube. We pick a particular vertex of all 16 vertices of the 4-cube and we obtain 15 vectors that point from this vertex to the remaining ones. The corresponding 15 central hyperplanes cut the unit 3-sphere as boundary of the four-dimensional unit ball in

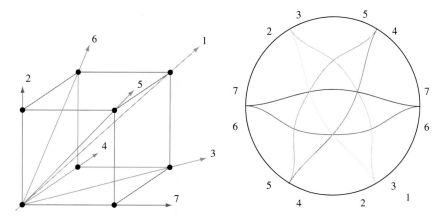

Figure 1.21 Contraction at a vertex of a 3-cube. Difference vectors on the left and representation of the sphere system on the right

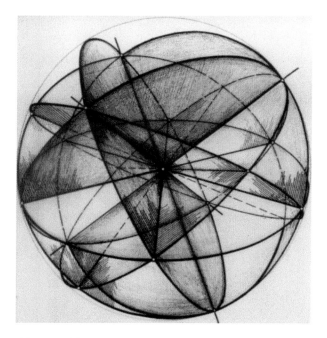

Figure 1.22 Combinatorial cube properties (drawing)

15 two-dimensional great spheres. We pick one of the 15 central hyperplanes and we project one half of the boundary structure of that ball onto this hyperplane. We obtain a three-dimensional ball with images under this projection of its two-dimensional former great spheres (in 14 cases we have just one half of them). In Figure 1.24 we see a corresponding model that shows the intersections of all

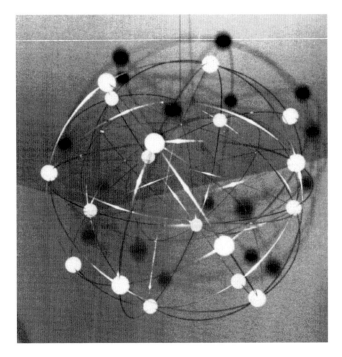

Figure 1.23 Combinatorial cube properties (model). Half of this boundary of a 3-ball, projected onto a plane as a disc, should be compared with Figure 1.24, a 3-ball, image of half of a projected boundary of a 4-ball

Figure 1.24 Combinatorial cube properties, dimension 4. Photo: M. Las Vergnas

Figure 1.25 Contraction at a vertex of an icosahedron, pottery model

pairs of 2-spheres under this projection from 4-space onto the three-dimensional hyperplane in our ordinary space. The model was a gift of the author to Michel Las Vergnas who took the photo of the model in his garden.

We present another object as a pottery model in Figure 1.25. It shows, in the same way as we have seen in the contraction at a vertex of a cube, the contraction at a vertex of an icosahedron.

It seems that this way of looking at a familiar object has no advantage at all; one reason for this might be that we have no experience so far. We will see in the next chapter that these representations are very natural for our generalizations of matrix equivalence classes, that is, for oriented matroids.

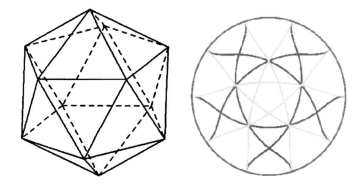

Figure 1.26 Icosahedron, on the left: for a fixed vertex we have 11 difference vectors pointing to all remaining vertices. They determine 11 central planes that determine the sphere system in Figure 1.25. On the right: the projection of the sphere system onto the plane has been drawn as a homeomorphic picture

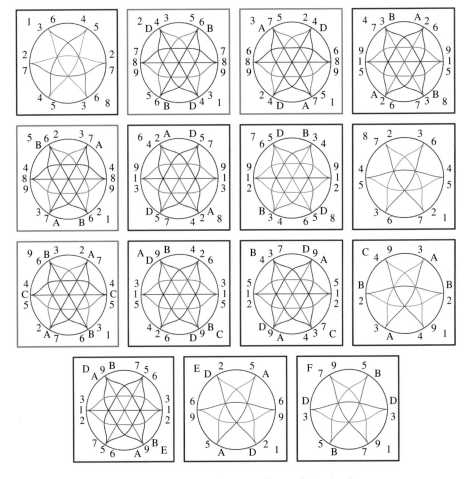

Figure 1.27 Rank 3 contractions of the 4-cube

Figure 1.28 Rank 3 contractions of the 24-cell

Finally, as a demonstration that our tools go beyond easy applications, we show in Figure 1.27 all projective plane sections of the 4-cube contraction model of Figure 1.24. These pictures result from the omawin software.

In the same manner we can study a contraction of the 24-cell in Figure 1.28.

2

Geometric matrix models II

2.1 A vector space and its dual

The r columns of our matrix define vectors in $\mathbb{R}^n$. These vectors (we assume that they are linearly independent) generate an r-dimensional *vector subspace* within $\mathbb{R}^n$. The normalized representation MR of our example matrix can be interpreted as a three-dimensional vector space in $\mathbb{R}^5$. The two bases of this three-dimensional vector space differ just by a basis transformation.

$$
MR := \begin{pmatrix} 0 & -1 & 1 \\ 0 & 0 & 1 \\ -1 & -1 & -1 \\ 0 & -1 & -1 \\ 1 & 1 & 0 \end{pmatrix} \begin{pmatrix} +1 & -2 & -1 \\ -1 & 1 & 0 \\ 0 & 1 & 0 \end{pmatrix} = \begin{pmatrix} 1 & 0 & 0 \\ 0 & 1 & 0 \\ 0 & 0 & 1 \\ 1 & -2 & 0 \\ 0 & -1 & -1 \end{pmatrix}.
$$

The vector space V has a well-defined *orthogonal space* V^* in $\mathbb{R}^n$, $n = 5$. We might as well assume that our matrix M was written to represent that space.

Recall that the chirotope information in the former section can be obtained from the signs of determinants of *all* submatrices of the Tucker matrix T of MR

$$
T := \begin{pmatrix} t_{1,1} & t_{1,2} & t_{1,3} \\ t_{2,1} & t_{2,2} & t_{2,3} \end{pmatrix} = \begin{pmatrix} 1 & -2 & 0 \\ 0 & -1 & -1 \end{pmatrix}.
$$

The same information is also contained in the negative transpose of the matrix T. This in turn leads to the signs of determinants of 2×2 submatrices of the following matrix.

$$
\begin{pmatrix} -t_{1,1} & -t_{2,1} \\ -t_{1,2} & -t_{2,2} \\ -t_{1,3} & -t_{2,3} \\ 1 & 0 \\ 0 & 1 \end{pmatrix} = \begin{pmatrix} -1 & 0 \\ 2 & 1 \\ 0 & 1 \\ 1 & 0 \\ 0 & 1 \end{pmatrix}.
$$

These two column vectors are orthogonal to all three column vectors of our matrix M. In other words, these two column vectors span the dual vector space V^*. Our matrix M can be viewed to represent just one part of the complete picture depicted in the following (5×5) matrix.

$$
\begin{pmatrix}
1 & 0 & 0 & -t_{1,1} & -t_{2,1} \\
0 & 1 & 0 & -t_{1,2} & -t_{2,2} \\
0 & 0 & 1 & -t_{1,3} & -t_{2,3} \\
t_{1,1} & t_{1,2} & t_{1,3} & 1 & 0 \\
t_{2,1} & t_{2,2} & t_{2,3} & 0 & 1
\end{pmatrix}.
$$

It is useful to look at the chirotope information again and to compare the chirotope of the original matrix with what we obtain from the matrix

$$
\begin{pmatrix}
-t_{1,1} & -t_{2,1} \\
-t_{1,2} & -t_{2,2} \\
-t_{1,3} & -t_{2,3} \\
1 & 0 \\
0 & 1
\end{pmatrix}
=
\begin{pmatrix}
-1 & 0 \\
2 & 1 \\
0 & 1 \\
1 & 0 \\
0 & 1
\end{pmatrix}.
$$

Since the Tucker matrix forms the link between both concepts, and the quadratic submatrices in both cases are just transposes of each other, it is easy to carry over the sign of the determinant information from the first chirotope to the other, the dual one.

The transition between both chirotopes can be described on a purely combinatorial basis as well. This gives us an impression of the important concept of duality for oriented matroids. The duality concept applied here (different from polar duality) is that from linear programming or that underlying the Gale diagram concept. Here we see a generalization of it.

Exercise: Write a Haskell program that calculates the dual chirotope from the primal one.

The following considerations should help to solve that exercise.

$$
\begin{array}{c|ccccccccc|c}
1 & 1 & 0 & \cdots & 0 & -11 & -21 & \cdots & -(n-r)1 & & 1 \\
2 & 0 & 1 & \cdots & 0 & -12 & -22 & \cdots & -(n-r)2 & & 2 \\
\vdots & \vdots & \vdots & \ddots & \vdots & \vdots & \vdots & \ddots & \vdots & & \vdots \\
r & 0 & 0 & \cdots & 1 & -1r & -2r & \cdots & -(n-r)r & & r \\
r+1 & 11 & 12 & \cdots & 1r & 1 & 0 & \cdots & 0 & & r+1 \\
r+2 & 21 & 22 & \cdots & 2r & 0 & 1 & \cdots & 0 & & r+2 \\
\vdots & \vdots & \vdots & & \vdots & \vdots & \vdots & \ddots & \vdots & & \vdots \\
n & (n-r)1 & (n-r)2 & \cdots & (n-r)r & 0 & 0 & \cdots & 1 & & n.
\end{array}
$$

The above matrix scheme, in which we assume $r \leq n - 1$, can be used to understand the relationship between a chirotope and its dual one. The Tucker matrix and the negative transpose of it appear together with two unit matrices. Note that the scalar product is zero for each of the first r column vectors with each of the remaining $n - r$ column vectors of the full matrix. In other words, the vector spaces spanned by these two sets of vectors are orthogonal. This duality is that from linear programming. If we study how the chirotope information of the first matrix can be carried over to the second, we will see that the resulting transformation is purely combinatorial in nature. This leads to the duality concept for oriented matroids that we are going to explain in more detail.

Let us consider an arbitrary ordered r-tuple $(j_1, j_2, \ldots, j_r)$ of row indices with $1 \leq j_1 < j_2 < \cdots < j_r \leq n$. Let $q \in \{0, 1, 2, \ldots, \min(r, n - r)\}$ be the number of those row indices j_α in the r-tuple $(j_1, j_2, \ldots, j_r)$ with $r + 1 \leq j_\alpha \leq n$. By applying the Laplacian rule for determinants, we see that the determinant of the chirotope of M corresponding to the tuple $(j_1, j_2, \ldots, j_r)$ is equal (the sign has still to be worked out) to the determinant of a $(q \times q)$-submatrix of the Tucker matrix. We obtain this submatrix from the Tucker matrix by keeping the q rows corresponding to row indices j_α with $r + 1 \leq j_\alpha \leq n$ and by deleting $r - q$ columns of the Tucker matrix. We delete column j_α, if the row index j_α in the r-tuple $(j_1, j_2, \ldots, j_r)$ lies in the closed interval from 1 to r. Now we look at the right $n - r$ columns of the full matrix. We observe that the complement of the set $\{j_1, j_2, \ldots, j_r\}$ with respect to the set $\{1, 2, \ldots, n\}$, ordered in the natural way and defining in this way the dual tuple $\{k_1, k_2, \ldots, k_{n-r}\}$ plays the decisive counterpart. This means, so far up to a sign argument, the following. The tuple $\{k_1, k_2, \ldots, k_{n-r}\}$, applied to the right $n - r$ columns of the full matrix, leads to the determinant of the same (transposed) submatrix of the Tucker matrix as the set $\{j_1, j_2, \ldots, j_r\}$ does for the original matrix.

If we count the total number of signs of the chirotope on the one hand with respect to the original matrix and on the other hand with respect to the Tucker matrix, we obtain a well-known relation between binomial coefficients:

$$\sum_{q=0}^{\min(r, n-r)} \binom{n-r}{r-q} \binom{r}{q} = \binom{n}{r}.$$

We understand now, up to the more technical sign argument in each particular case, how the $\binom{n}{r}$ many signs of the original chirotope can be carried over via the Tucker matrix to define corresponding $\binom{n}{n-r}$ signs of the chirotope of the right part of the matrix, the dual chirotope. But a second look at this technical machinery shows that once we know how the signs carry over to the new chirotope, we can forget the matrix concept. The matrix considerations only indicated what to do. In this way we have introduced the duality concept for oriented matroids.

We now discuss the more technical part of how the sign of the r-tuple $(j_1, j_2, \ldots, j_{r-q}, j_{r-q+1}, \ldots, j_r)$ has to be converted into the sign of the complementary $(n-r)$-tuple $(k_1, k_2, \ldots, k_q, k_{q+1}, k_{q+2}, \ldots, k_{n-r})$. We have used a notation such that $1 \leq j_1 < j_2 < \cdots < j_{r-q} \leq r < j_{r-q+1} < \cdots < j_r \leq n$ and $1 \leq k_1 < k_2 < \cdots < k_q \leq r < k_{q+1} < \cdots < k_r \leq n$ and

$$\{j_1, j_2, \ldots, j_r\} \cup \{k_1, k_2, \ldots, k_{n-r}\} = \{1, 2, \ldots n\}.$$

We interchange adjacent column vectors of the left matrix until the number one in row number j_1 has its position in the first column, we use $j_1 - 1$ changes for that. We proceed to make similar adjacent transpositions until the number one in row number j_2 has its position in the second column, etc. After q such steps, we see that the sign of the tuple $(j_1, j_2, \ldots, j_r)$ has to be multiplied with $(-1)^{j_1-1}(-1)^{j_2-2}\ldots(-1)^{j_{r-q}-(r-q)}$ in order to fit with the sign of the corresponding determinant of the $(q \times q)$-submatrix of the Tucker-matrix.

For the right submatrix, with row indices $k_1, k_2, \ldots, k_{n-r}$, we also aim to use column and row transpositions such that the normalized version of the submatrix of the transposed Tucker matrix appears. We look at the number one in row k_{q+1}. After $k_{q+1} - (r+1)$ changes of adjacent columns, its position is in the $(r+1)$th column of the full matrix, etc. Our sign has to be corrected by the factor

$$(-1)^{k_{q+1}-(r+1)}(-1)^{k_{q+2}-(r+2)}\ldots(-1)^{k_{n-r}-(2r-q)}.$$

Now we use $q(r-q)$ transpositions of rows to move the $(n-r-q) \times (n-r-q)$ unit matrix to the top and we take into account that we now have the negative transpose Tucker matrix, which yields the factor $(-1)^q$ and we obtain altogether the factor

$$(-1)^{\sum_{i=1}^{r-q}(j_i-i)+\sum_{i=1}^{n-r-q}(k_{q+i}-(r+i))+q(n-r-q)+q}$$

between the signs of the tuples $(j_1, j_2, \ldots, j_r)$ and $(k_1, k_2, \ldots, k_{n-r})$. This can now be written as some Haskell code.

How to obtain the dual chirotope of a chirotope

```
dC::Int->Int->[Int]->[Int]
dC r n l=reverse[(l!!i)*(plm(mlexp n r ((tuples r n)!!i)))
                              |i<-[0..length(tuples r n)-1]]
mlexp::Int->Int->[Int]->Int
mlexp n r tuple
 | q == 0   = 0
 | otherwise = q*(n-r-q + 1)
              +(sum[(tuple!!(i-1))-i | i<-[1..(r-q)]])
              +(sum[(([1..n]\\ tuple)!!(q+i-1))-(r + i)
                       | i<-[1..(n-r-q)] ])
                    where q = tu2q r n tuple
tu2q::Int->Int->[Int]->Int
tu2q r n [] = 0
tu2q r n (h:li)|h 'elem'[(r+1)..n]=1+tu2q r n li
               |otherwise         = tu2q r n li
plm::Int->Int
plm n |n==0 = 1 |n==1 = -1 |otherwise= plm(n-2)
```

The function dC uses the rank r, the number of elements n, and the list of orientations of the chirotope as input and returns the list of orientations of the dual chirotope with n elements of rank $n - r$.

We can use the alternating oriented matroid from Page 15 as an example. We obtain from dC 3 6 (map snd (altOM 3 6)) the result

$$[-1, 1, -1, 1, -1, 1, -1, -1, 1, -1, 1, -1, 1, 1, -1, 1, -1, 1, -1, 1]$$

and we can understand the name for this chirotope. Try also dC 5 8 (map snd (altOM 5 8)).

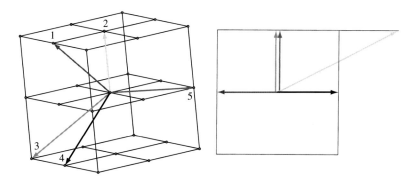

Figure 2.1 Corresponding vectors in V and V^*

If we look at this duality in the theory of oriented matroids from a purely combinatorial point of view, both the former concepts benefit from this understanding. We can clearly distinguish between metrical properties and combinatorial properties. Duality for chirotopes is combinatorial in nature.

2.2 A point on a Grassmannian

The essential information of the dual pairs of the last section was stored in the Tucker matrix. In our example matrix M this means that we were looking at just six parameters. This way of looking at our problem leads to a manifold structure with six independent coordinates.

The determinants of all $(r \times r)$-submatrices of an $(n \times r)$-matrix are not independent, since the same entries occur in several of them. The dependences are known as Grassmann–Plücker relations.

Let $E := \{1, 2, \ldots, n\}$ be our set of elements with the natural order and let $r \leq n$ be a natural number. We define the ordered tuples of r-element subsets of E as $\Lambda(n, r) := \{(\lambda_1, \ldots, \lambda_r) \in E^r | 1 \leq \lambda_1 < \cdots < \lambda_r \leq n\}$. For $\lambda = (\lambda_1, \ldots, \lambda_r) \in E^r$ we consider a formal variable $[\lambda_1, \ldots, \lambda_r]$. We require for any permutation of r elements $\pi \in S_r$

$$[\pi(\lambda_1), \ldots, \pi(\lambda_r)] := \text{sign}(\pi) [\lambda_1, \ldots, \lambda_r].$$

This fixes the values of all r-tuples if the values on the set $\Lambda(n, r)$ are defined.

We define *k-term Grassmann–Plücker relations* (GP relations) as follows. For $3 \leq k \leq r+1$ and three sets of indices $A := \{a_1, \ldots, a_{r-k+1}\}$, $B := \{b_1, \ldots, b_{k-2}\}$, and $C := \{c_1, \ldots, c_k\}$ with pairwise different elements from E, we define the polynomial

$$\{A|B|C\} := \sum_{i=1}^{k} (-1)^{i+1} \alpha(A, C, i) \cdot \beta(A, B, C, i)$$

$$\alpha(A, C, i) := [a_1, \ldots, a_{r-k+1}, c_1, \ldots, \hat{c}_i, \ldots, c_k]$$

$$\beta(A, B, C, i) := [a_1, \ldots, a_{r-k+1}, b_1, \ldots, b_{k-2}, c_i]$$

We use the notation $\hat{c}_i$ to indicate that c_i has been discarded. We call the polynomial equation $\{A|B|C\} = 0$ a *k-term Grassmann–Plücker relation*. The notation $\{A|B|C\} = 0$ was chosen since any permutation of elements within one of the sets A, B, C does not change the Grassmann–Plücker relation.

For $k = 3$ there is still a redundancy in the above formula that we are going to reduce for algorithmical purposes since 3-term Grassmann–Plücker relations play a special role. We define these relations for two sets $A := \{a_1, \ldots, a_{r-2}\}$, $B := \{b_1, \ldots, b_4\}$ with $A, B \subset E$ and $A \cap B = \emptyset$ in the form

$$\{a_1, \ldots, a_{r-2} | b_1, \ldots, b_4\}$$

$$:= +[a_1, \ldots, a_{r-2}, b_1, b_2][a_1, \ldots, a_{r-2}, b_3, b_4]$$

$$-[a_1, \ldots, a_{r-2}, b_1, b_3][a_1, \ldots, a_{r-2}, b_2, b_4]$$

$$+[a_1, \ldots, a_{r-2}, b_1, b_4][a_1, \ldots, a_{r-2}, b_2, b_3] = 0.$$

Again, this relation is invariant with respect to permutations within the sets A and B. For a given $(n \times r)$-matrix with row_i as its ith row we define

$$[\lambda_1, \ldots, \lambda_r] := \det(\text{row}_{\lambda_1}, \ldots, \text{row}_{\lambda_r}).$$

Grassmann–Plücker relations, general case

```
kTermGPrels::Int->Int->Int->[[([Int],[Int])]]
kTermGPrels k r n
 |k/=3=[ map(i->(a++(take (i-1) c)++(drop i c),
                 a++b++(take 1(drop (i-1)c))))[1..k]
       |a<-tuplesL (r-k+1)[1..n],b<-tuplesL(k-2)([1..n]\\ a),
        c<-tuplesL k ([1..n]\\(a++b))]
 |k==3=[[(a++[b!!0]++[b!!1], a++[b!!2]++[b!!3])]
       ++[(a++[b!!0]++[b!!2], a++[b!!1]++[b!!3])]
       ++[(a++[b!!0]++[b!!3], a++[b!!1]++[b!!2])]
       |a<-tuplesL(r-2)[1..n],b<-tuplesL 4([1..n]\\ a)]
```

Before we prove these relations, we have written some Haskell code that can generate all of them. We observe again that the program is very close to its mathematical description.

Proof of 3-term GP-relations for a matrix

We consider the submatrix of rows $a_1, \ldots, a_{r-2}, b_1, \ldots, b_4$, we relabel the rows as $1, 2, \ldots, r+2$, and we assume without loss of generality that the first r row vectors are linearly independent. This allows us, by multiplying the whole matrix with the inverse submatrix, to write the unit matrix for the first r row vectors.

$$
\begin{array}{c}
1 \\
2 \\
\vdots \\
r-1 \\
r \\
r+1 \\
r+2
\end{array}
\left(
\begin{array}{cccccc}
1 & 0 & \ldots & 0 & 0 \\
0 & 1 & \ldots & 0 & 0 \\
\vdots & \vdots & \ddots & \vdots & \vdots \\
0 & 0 & \ldots & 1 & 0 \\
0 & 0 & \ldots & 0 & 1 \\
\cdot & \cdot & \ldots & a & b \\
\cdot & \cdot & \ldots & c & d
\end{array}
\right).
$$

The 3-term GP-relation turns out to be Laplacian's rule for a (2×2) determinant.

$$
\{1, \ldots, r-2 \mid r-1, r, r+1, r+2\}
$$
$$
:= +[1, \ldots, r-2, r-1, r] \cdot [1, \ldots, r-2, r+1, r+2]
$$
$$
- [1, \ldots, r-2, r-1, r+1] \cdot [1, \ldots, r-2, r, r+2]
$$
$$
+ [1, \ldots, r-2, r-1, r+2] \cdot [1, \ldots, r-2, r, r+1]
$$
$$
= 1 \cdot (ad - bc) - b \cdot (-c) + d \cdot (-a) = 0.
$$

Idea for proving a k-term GP-relation

The k-term GP-relation for a matrix holds as well. The reason for this is again Laplacian's rule for determinants. Write a unit $(r \times r)$-matrix on top of a $(r \times r)$-matrix and calculate the product of both corresponding determinants by evaluating the second determinant according to the Laplacian rule, say along its first row. The resulting homogeneous relation is a GP relation:

For any set of indices $a_1, \ldots, a_{r+1}, b_1, \ldots, b_{r-1} \in \{1, \ldots, n\}$, we have

$$
\sum_{i=1}^{r+1} (-1)^i [a_1, \ldots, \hat{a}_i, \ldots, a_{r+1}][a_i, b_1, \ldots, b_{r-1}] = 0.
$$

By interpreting these equations as the intersection of hypersurfaces in a space of dimension $\mathbb{R}^{\binom{n}{r}}$ in which the brackets are the coordinates, we have defined *the Grassmann cone*. When we consider the intersection of this cone with a unit sphere, we arrive at the Grassmannian. Each $(n \times r)$-matrix defines a point on this manifold and up to a non-singular transformation, this matrix is uniquely determined by this point on the Grassmannian.

We come back to the idea of studying invariants under transformation groups proposed by Felix Klein in his Erlanger Program. By replacing the matrix under consideration by its point on a Grassmannian, we already have an object which represents an equivalence class of matrices, that is, those which coincide up to a multiplication by a non-singular $(r \times r)$-matrix or, in other words, those which are invariant under the orthogonal group $O(r)$.

The foregoing models of a matrix and the models still to come can be considered without fixing the order of its elements and without fixing its orientations. We are sometimes not interested in the orientations of the lines, planes, circles, etc. in which cases we consider the so-called *reorientation class*, that is the equivalence class of our objects up to reorientating a subset of its elements. In addition, we sometimes consider another equivalence relation, an invariant with respect to permutations within the set E of our elements.

Geometric interpretation of a Grassmann–Plücker relation

Figure 2.2 shows the geometric picture of a 3-term Grassmann–Plücker relation that provides a link between this relation and hyperline sequences.

We have depicted the affine hull of the points $A1, A2, \ldots, A(r-2)$ and the additional four points $B1, B2, B3, B4$ that lie in this case all on the same side with respect to affine hull of $A1, A2, \ldots, A(r-2), B1$. The sign condition that we will deduce later from the 3-term GP-relation asserts that the orientation of the arrow from $B2$ to $B3$ and the orientation of the arrow from $B3$ to $B4$ implies the orientation of the arrow from $B2$ to $B4$ in the circular sequence around the hyperline.

For a general oriented matroid definition, the chirotope version, we will consider a map that assigns for each abstract simplex of r elements an orientation. We will require that these orientations, if interpreted formally as signs of determinants, do not contradict any 3-term Grassmann–Plücker relation. It is useful in this context to have the geometric picture of Figure 2.2 in mind.

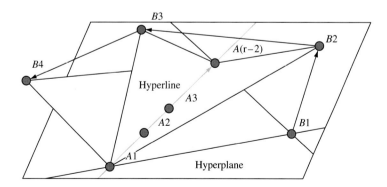

Figure 2.2 Three-term Grassmann–Plücker relation and hyperline idea

2.3   A cell decomposition in projective space

The oriented circles on the sphere, that we have used in Section 1.5, define a
point symmetric cell decomposition of the unit sphere in $\mathbb{R}^3$ (with respect to the
midpoint of the sphere). By identifying antipodal points on the unit sphere, we
obtain the *projective plane*. The given matrix can be interpreted as defining a *cell
decomposition of the projective plane*.

The cell decomposition of the sphere presented in Section 1.5 is symmetric
with respect to the midpoint. If we identify a point with its antipodal point, the
cell decomposition on the sphere with its double covering of the projective plane
defines a cell decomposition in the corresponding projective space, in our example
in the projective plane. This information is contained in an orthogonal projection
of the upper half, or the lower half, of the sphere onto a central hyperplane.
We prefer to use a projection such that one element forms the boundary of
the corresponding $(r-1)$-ball. Usually, we use only one picture out of the five
possible equivalent ways to represent the cell decomposition. In Figure 2.3 we
see all of them. We have always depicted the negative hemisphere corresponding
to the orientation of the chosen boundary element.

The p-vector of the cell decomposition of the projective plane collects the
number of triangles, quadrangles, pentagons, etc. as its components. It is an
example of a concept that is invariant under forming reorientations on subsets of
E and under permutations on the ground set E. We see eight triangles and one
quadrangle in the five versions of rendering our cell structure in Figure 2.3. The
quadrangle has always been depicted in blue. So the p-vector in our example
is $p = (8, 1, 0, \ldots)$.

The cell decomposition of the projective plane can be generated for $(n \times 3)$-
matrices automatically with the program omawin, oriented matroid manipulator
using x-windows. In fact, Figure 2.3 already has shown such a picture. As
input file we use the chirotope information in the file "exemplify.chi." This file
contains the following: we have five row elements, the number of columns of the
matrix is three and the signs of the determinants of (3×3)-submatrices in their
natural order are contained in the last line. Hence the input file for Figure 2.3
is just:

Data input for omawin program

```
5
3
+0-++++0+-
```

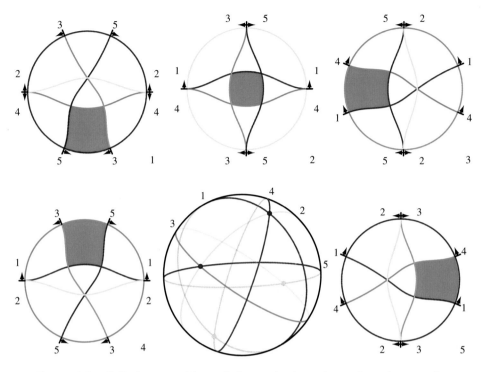

Figure 2.3 Cell decomposition of the projective plane given by our five elements and corresponding cell decomposition on the sphere

Recall that our chirotope was of the form

$$\chi(\mathbf{M}) := (+\mathbf{1}, \mathbf{0}, -\mathbf{1}, +\mathbf{1}, +\mathbf{1}, +\mathbf{1}, +\mathbf{1}, \mathbf{0}, +\mathbf{1}, -\mathbf{1}).$$

The output of Figure 2.3 can be generated by the call ./omawin and by using the input file "exemplify.chi." The view option allows us to rotate the picture and to change the *line at infinity*, that is, the element which is used as the bounding circle on which antipodal points have to be identified. We use another file for a septagon:

Data input for omawin program

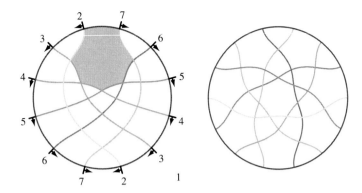

Figure 2.4   Cell decomposition corresponding to a seven-gon

The omawin program changes this into the left part of Figure 2.4. If we insert an additional element as the line at infinity, we obtain the seven-gon in the way that we would expect, see the right part of Figure 2.4.

2.4   Topes, covectors, cocircuits, big face lattice

We can define the blue cell in Figure 2.3 either by the sign vector

$$
\begin{array}{c}
1\ 2\ 3\ 4\ 5 \\
(-,-,+,+,+)
\end{array}
$$

or by its sign reversal

$$
\begin{array}{c}
1\ 2\ 3\ 4\ 5 \\
(+,+,-,-,-)
\end{array}
$$

We obtain this signed vector by picking a relative interior point in a cell and by deciding for each oriented circle on which side the point lies. Opposite points on the sphere correspond to a sign reversal of all elements. The four edges of this four-gon can be described by the following vectors (on the left) or by their sign reversals. The position of the zero corresponds to the element on which this line segment lies.

$$
\begin{array}{c}
1\ 2\ 3\ 4\ 5 \\
(0,-,+,+,+) \\
(-,-,0,+,+) \\
(-,-,+,0,+) \\
(-,-,+,+,0)
\end{array}
\qquad
\begin{array}{c}
1\ 2\ 3\ 4\ 5 \\
(0,-,0,+,+) \\
(-,-,0,0,+) \\
(-,-,+,0,0) \\
(0,-,+,+,0)
\end{array}
$$

The four vertices correspond to the sign vectors on the right. The position of the zero indicates which elements intersect in that particular vertex. In these four vertices the minimal number of elements intersect.

We have described all faces (the face lattice) of the four-gon by sign vectors. We see from the picture that we can add sign vectors of the remaining zero-dimensional, one-dimensional and two-dimensional cells. All these sign vectors are determined by the given matrix. The collection of all these sign vectors represent this big face lattice. The partial order defining the face lattice is clear from the construction.

We present the big face lattice in terms of all signed vectors (covectors) for our example matrix M. We omit the reversed sign vectors. Compare this with Figure 2.3 and recall that these sign vectors were given by our example matrix. In other words, the matrix can be assumed to represent this information. We have sign vectors representing the cells of maximal dimension, the so-called *topes*. We have nine of them up to reorientation. Sign vectors representing zero dimensional cells (we have six of them up to reorientation) are called *cocircuits*. The general notion for such a sign vector is *covector*.

```
1 2 3 4 5   1 2 3 4 5   1 2 3 4 5   1 2 3 4 5   1 2 3 4 5
(-,-,+,+,+) (-,-,+,+,-) (-,-,+,-,+) (-,-,+,-,-) (-,-,-,-,+)
(-,-,-,+,+) (-,+,+,-,+) (-,+,-,-,-) (-,+,-,-,+) 9 2-cells

(0,-,+,+,+) (0,-,-,+,+) (0,+,-,-,+)
(-,0,-,-,+) (-,0,+,-,-)
(-,-,0,+,+) (-,-,0,-,+) (-,+,0,-,+)
(-,-,+,0,-) (-,-,+,0,+) (-,-,-,0,+)
(-,-,+,-,0) (-,-,+,+,0) (-,+,+,+,0)            14 1-cells

(0,-,+,+,0) (0,-,0,+,+) (0,0,-,0,+)
(-,0,0,-,0) (-,-,0,0,+) (-,-,+,0,0)            6 0-cells
```

2.5 A zonotope, Minkowski sum of line segments

If we stress the fact that the vectors corresponding to the rows of our given matrix M define line segments, we can use the Minkowski sum of these line segments in order to obtain another model, the corresponding *zonotope*.

The rows of the matrix actually contain the information of oriented line segments. So, we can use this information and we can draw the zonotope defined by such oriented line segments. If we forget the orientations of our line segments, we have a model for an equivalence class of matrices, those which differ by just replacing certain rows with their negative ones. In other words, this information,

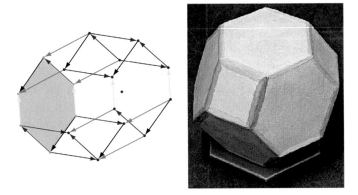

Figure 2.5   Zonotope that corresponds to our example Matrix M, computer graphics (left); Minkowski sum of the edges of a regular tetrahedron, pottery model (right)

the reorientation class of matrices given by our matrix, is represented by the zonotope (with unoriented edges).

If K_i $i \in \{1, \dots, n\}$ is a finite set of convex sets in a Euclidean space, we define the Minkowski sum as $K_1 + \cdots + K_n := \{x \mid x = a_1 + \cdots + a_n, \, a_i \in K_i\}$. Dependences within the set of vectors appear as degenerated facets. Our example, Figure 2.5 left, can also be viewed as the Minkowski sum of two hexagons having a common edge.

2.6   A projection of a cube

A zonotope can be viewed as a parallel projection of a regular cube from dimension n onto r-space. The remaining $n - r$ columns of the matrix have been discarded. The equivalence of zonotopes with projections of cubes is well known. The permutahedron is a nice example in this context.

Consider the permutations of four elements $\{1, 2, 3, 4\}$ and interpret each permutation $(\pi(1), \pi(2), \pi(3), \pi(4))$, $\pi \in S_4$, as a point in Euclidean 4-space.

The convex hull of these points lies in a hyperplane since the sum over all components is constant. It turns out via an easy exercise using linear algebra arguments that this convex hull is one of the Archimedean bodies with squares and regular hexagons as facets. The permutahedron is the projection of the 6-cube onto Euclidean 3-space. It is the Minkowski sum of the edges of a regular tetrahedron, see Figure 2.5 (right) and Figure 2.7.

In Figure 2.6 we have drawn the projection of some part of the four-dimensional unit lattice. We have marked the vertices of the permutahedron. In this way, the permutahedron can be easily seen.

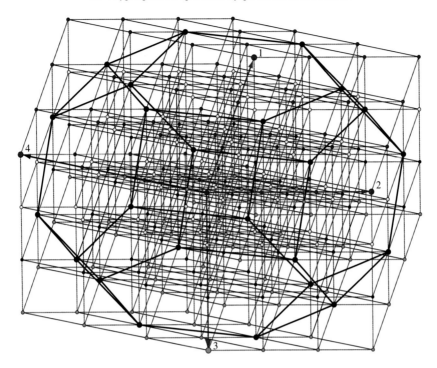

Figure 2.6   Permutahedron in the four-dimensional lattice

2.7   Hyperplanes spanned by point sets, cocircuits

We look again at Figure 1.6 or Figure 1.17 with the given 5 row vectors.

```
1  2  3  4  5        1  2  3  4  5
(0,0,-,0,+)        (0,0,+,0,-)
(0,+,0,-,-)        (0,-,0,+,+)
(0,+,-,-,0)        (0,-,+,+,0)
(+,0,0,+,0)        (-,0,0,-,0)
(+,+,0,0,-)        (-,-,0,0,+)
(+,+,-,0,0)        (-,-,+,0,0)
```

Any two vectors span a central plane and the remaining vectors are split into three sets, those lying in the plane and those lying in the positive halfspace or in the negative one. We write this down in the form of signed vectors to indicate these cases. These signed vectors are called *cocircuits* and the set of all cocircuits in our example case is the following.

This is more information that we can inherit from our given matrix M. We carry over our rule to the affine case. The cocircuits can be interpreted as abstract

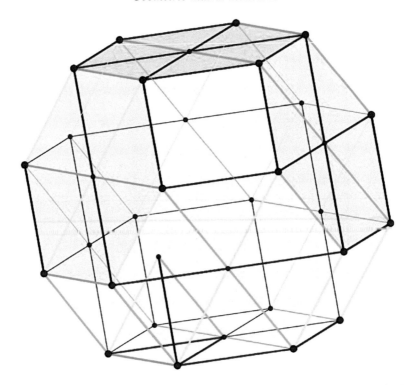

Figure 2.7　Edge analysis of the permutahedron with corresponding tetrahedron

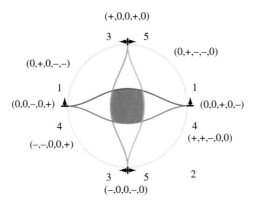

Figure 2.8　Sphere system with covectors

lines (affine hyperplanes in higher dimensions) in the affine plane (hyperplane). They tell us the point sets that are separated by the corresponding line. If we carry over the rule to the affine case, the signed points have to be taken into account.

2.8 Minimal dependent sets, circuits

We call a subset of n row vectors a minimal dependent set if it is linearly dependent and if all proper subsets are linearly independent. In the case of our example matrix M with 5 rows and indices $\{1, 2, 3, 4, 5\}$, we have minimal dependent sets with three and four elements as follows:

$$\{1, 2, 4\}, \ \{2, 3, 5\}, \ \{1, 3, 4, 5\}.$$

Here is some Haskell code that can find this information:

We provide first two functions that are not part of the Haskell prelude file, the vector product in three-dimensions and a subset function. Both functions need no further explanation.

Vector product and subset function

```
xProd::MA->[Integer]
xProd [v1,v2] =[ det [[v1!!1,v1!!2],[v2!!1,v2!!2]],
                -(det [[v1!!0,v1!!2],[v2!!0,v2!!2]]),
                 det [[v1!!0,v1!!1],[v2!!0,v2!!1]] ]

isSubSet::[Int]->[Int]->Bool
isSubSet []       set = True
isSubSet (x:xs) set | x 'elem' set = isSubSet xs set
                    | otherwise   = False
```

Minimal dependent sets of our matrix M

```
minDep2::MA->[[Int]]
minDep2 m
 =[[a,b]|[a,b]<-tuples 2 5,xProd[m!!(a-1),m!!(b-1)]==[0,0,0]]

minDep3::MA->[[Int]]
minDep3 m=[t|t<-tuples 3 5,and[not(isSubSet p t)|p<-mD],
           head(dets [t] m)==0]
                              where mD = minDep2 m
minDep4::MA->[[Int]]
minDep4 m=[q|q<-tuples 4 5,and[not(isSubSet t q)|t<-mD]]
          where mD = minDep2 m ++ minDep3 m
```

With `union (minDep4 m) (union (minDep2 m) (minDep3 m))` we obtain the result `[[1,3,4,5],[1,2,4],[2,3,5]]` of all minimal dependent

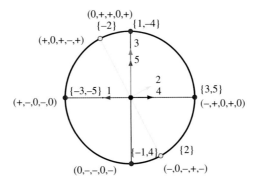

Figure 2.9   Circuits marked in the dual sphere system

sets in our case. Apart from prelude functions and functions of the list module, we have used the functions `minDep2`, `minDep3`, `minDep4`, `isSubSet`, `tuples`, `xProd`, `dets`, and `det`.

For a minimal dependent set of k vectors v_i, there is a relation $\sum_{i=1}^{k} \lambda_i v_i = 0$ with non-zero coefficents λ_i. The coefficients λ_i are uniquely defined up to a non-zero factor. Here are the dependences in our case with their coefficients. The three expresssions in the following lines are equal to the zero vector.

```
1*[0,-1,1]+(-2)*[0,0,1]+0*[-1,-1,-1]+(-1)*[0,-1,-1]+0*[1,1,0]

0*[0,-1,1]+  1 *[0,0,1]+1*[-1,-1,-1]+  0 *[0,-1,-1]+1*[1,1,0]

1*[0,-1,1]+  0 *[0,0,1]+2*[-1,-1,-1]+(-1)*[0,-1,-1]+2*[1,1,0]
```

The signs of these coefficients can be used to define a partition of the set of k vectors into two subsets A and B. These partitions can be encoded as a pair of signed vectors of length n. The vectors of the pair are equal up to reversing all signs. We obtain such a vector by representing the ith component by 0 if the ith row is not within the minimal dependent subset and by $+/-$ if the ith row is in set A/B respectively.

We find the so-called *Radon partitions* in this way, that is, the positive convex cone generated by the vectors with positive sign and the positive cone generated by the vectors with negative sign have precisely one ray in common.

```
1  2  3  4  5       1  2  3  4  5
(+,-,0,-,0)    (-,+,0,+,0)
(+,0,+,-,+)    (-,0,-,+,-)
(0,+,+,0,+)    (0,-,-,0,-)
```

Circuits of a chirotope in rank 3

```
chi2C::[OB]->[[Int]]->[[Int]]->[[Int]]
chi2C chi [] cs = cs
chi2C chi (dep:deps) cs
 |length dep == 4
  = chi2C chi deps (cs++[[  cComp4 p chi dep |p<-sels]]
                    ++[[-1*(cComp4 p chi dep)|p<-sels]])
 |length dep == 3
  =chi2C chi deps (cs++[[  cComp3 p ch dep |p<-sels]]
                   ++[[-1*(cComp3 p ch dep)|p<-sels]])
 where  els = nub (concat (map fst chi))
        sels = [1..length els]
         ch  = ctrSetChi(take 1(els\\ dep))chi


cComp4::Int->[OB]->[Int]->Int
cComp4 p ch [a,b,c,z]
 |p==a=   msc!!(head(elemIndices[b,c,z]mfc))
 |p==b= -1*(msc!!(head(elemIndices[a,c,z]mfc)))
 |p==c=   msc!!(head(elemIndices[a,b,z]mfc))
 |p==z= -1*(msc!!(head(elemIndices[a,b,c]mfc)))
 |otherwise= 0    where msc = map snd ch
                       mfc = map fst ch
cComp3::Int->[OB]->[Int]->Int
cComp3  p ch [a,b,z]
 |p==a= -1*(msc!!(head(elemIndices [b,z] mfc)))
 |p==b=   msc!!(head(elemIndices [a,z] mfc))
 |p==z= -1*(msc!!(head(elemIndices [a,b] mfc)))
 |otherwise= 0    where msc = map snd ch
                       mfc = map fst ch
```

Again, we have determined a different set of signed vectors, called *circuits*, that came from our given matrix M.

Finding the values for the coefficient can be done via Cramer's rule that depends on only determinants. We have stored the signs of these determinants as the chirotope of the matrix. Hence it is natural to calculate the circuit information from the chirotope. We have done this as some Haskell code. Thinking of Cramer's rule and thinking of the contraction as an operation that comes from central projection helps us to understand the program. It has turned out that this calculation can be done on an abstract setting as well where no coordinates are involved.

```
> chi2C (altOM 6 3) (tuples 4 6) []
[[1,-1,1,-1,0,0],[-1,1,-1,1,0,0],[1,-1,1,0,-1,0],
 [-1,1,-1,0,1,0],[1,-1,1,0,0,-1],[-1,1,-1,0,0,1],
 [1,-1,0,1,-1,0],[-1,1,0,-1,1,0],[1,-1,0,1,0,-1],
 [-1,1,0,-1,0,1],[1,-1,0,0,1,-1],[-1,1,0,0,-1,1],
 [1,0,-1,1,-1,0],[-1,0,1,-1,1,0],[1,0,-1,1,0,-1],
 [-1,0,1,-1,0,1],[1,0,-1,0,1,-1],[-1,0,1,0,-1,1],
 [1,0,0,-1,1,-1],[-1,0,0,1,-1,1],[0,1,-1,1,-1,0],
 [0,-1,1,-1,1,0],[0,1,-1,1,0,-1],[0,-1,1,-1,0,1],
 [0,1,-1,0,1,-1],[0,-1,1,0,-1,1],[0,1,0,-1,1,-1],
 [0,-1,0,1,-1,1],[0,0,1,-1,1,-1],[0,0,-1,1,-1,1]]

> result
[(([1,2,3],1),([1,2,4],0),([1,2,5],-1),([1,3,4],1),
  ([1,3,5],1),([1,4,5],1),([2,3,4], 1),([2,3,5],0),
  ([2,4,5],1),([3,4,5],-1)]


> chi2C result((minDep3 result)++(minDep4 result))[]
[[1,-1,0,-1,0],[-1,1, 0,1, 0],[0,-1,-1, 0,-1],
 [0, 1,1, 0,1],[-1,0,-1,1,-1],[1, 0, 1,-1, 1]]     -}
```

The terminology of oriented matroid theory is often difficult to understand. Circuits and cocircuits have their origins in graph theory.

Graphs with their circuits and cocircuits

The remaining part of this section is devoted to graphs. A graph is a pair of a finite number of vertices and a set of two-element subsets of these vertices. We write the vertices of the graph as a sorted list, Vs = [Int] and we write the edges as a list of 2-element sublists of Vs, Es = [[Int]]. For the whole graph we use the type GR = (Vs,Es). See Page 303 for an overview of the chosen types for our data structures. A first easy example of a graph is the complete graph with *n* vertices. It has the following Haskell program.

Complete graph with *n* vertices

```
graphComplete::Int->([Int],[[Int]])
graphComplete n = ([1..n],tuples 2 n)
```

In the following we consider the graph to be oriented, that is, the ordering of the vertices of an edge is taken into account. A circuit in an oriented graph is a closed path in the graph that uses every edge at most once and that uses every vertex at most once. In Figure 2.10 we have drawn a graph (with parallel and anti-parallel

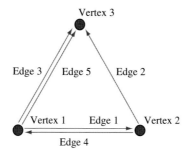

Figure 2.10   Circuits in a corresponding graph

edges) which has all circuits from our given example matrix (and more). The circuit function is not directly applicable later because of parallel edges.

In order to explain the terminology from graphs, we will work with an example which is not trivial; we use an oriented version of the Petersen graph, see Figure 2.11.

We look at the problem of finding all circuits in this Petersen graph. During the discussion of this example we will see graphs and vector spaces in context.

The Petersen graph can be obtained from the edge graph of the dodecahedron if we identify antipodal vertices and edges. We assume that the reader knows some basic terminology on graphs. The book of Bachem and Kern *Linear Programming Duality*, see Bachem and Kern, 1992, especially the second chapter, is a good additional source for the following considerations. A circuit in a graph is an alternating cyclic sequence of vertices and edges such that adjacent elements (vertices and edges) are incident and each element occurs precisely once.

We consider an oriented version of the Petersen graph. For its definition we use the following constant function `graphP`.

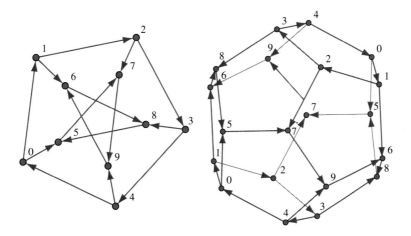

Figure 2.11   Petersen graph, an oriented version

Haskell program for the above Petersen graph

```
graphP::GR
graphP=([0..9], [ [a, ((a+1)'mod'5) ]       | a<-[0..4] ]
               ++[ [a, a+5 ]                 | a<-[0..4] ]
               ++[ [5+a, 5+((a+2)'mod'5)]| a<-[0..4] ])
```

We determine all circuits as two disjoint lists with `gr2C`. The first list contains all circuits with a fixed edge of the graph and a second list contains all circuits of the subgraph in which this edge has been deleted.

Circuits of an oriented graph

```
gr2C::GR->[[OE]]
gr2C (_,[]) = []
gr2C graph@(vs,(e:es))=(cEdge e graph)++gr2C(vs,es)

cEdge::[Int]->GR->[[OE]]
cEdge e@[z,a](vs,es)=map([(e,1)]++)(paths a
  z[](vs,es\\[e]))
```

The first set of circuits that we get from `cEdge` are paths that connect the endpoints of an edge that we have deleted. The problem remains to determine extended paths with given start and end points.

The function `paths` does the job.

Paths in an oriented graph

```
paths::Int->Int->[OE]->GR->[[OE]]
paths    i    z    path (vs,es)
=concat (map (\edge,or)->
(if z 'elem' edge then      [path++[(edge,or)]] else
paths(head(edge\\[i]))z(path++[(edge,or)])(vs,es\\[edge])))
orEdges)
where
fV = if path==[] then [] else tail (concat (map fst path))
orEdges =  [(([u,v], 1)|[u,v]<-es, u==i, v'notElem'fV ]
        ++ [([u,v],-1)|[u,v]<-es, v==i, u'notElem'fV ]
```

The old path is kept and extended by a new oriented edge in each step until the final point has been reached. Although the Petersen graph is 3-valent and repeated vertices cannot occur, we have written the function `paths` such that forbidden vertices are avoided in other example graphs. We obtain all circuits of our graph example by evaluating `gr2C graphP`.

The function `signVector` changes the representation of the circuits into signed vectors. We order all edges and we write their orientations as a corresponding vector. In the function, 0 at position i tells us that the corresponding ith edge is not used in the circuit, 1 and −1 gives us the orientation of that edge.

Sign vectors

```
signVector::[OE]->GR->[Char]
signVector  c  (vs,es)
=map(\e->(if(e,1)'elem'c then '+'
                    else(if(e,-1)'elem'c then '-'
                                else '0')))es
```

We obtain all circuits of the graph with

`resultCircuits=map(\oes->signVector oes graphP)(gr2C graphP)`

All circuits of our oriented version of the Peteresen graph

```
1234567 etc.    1234567 etc.    1234567 etc.    1234567 etc.
+++++0000000000 ++++0-000+0+0++ ++++0-000+-0-00 +++0+000+-+0++0
+++00-00+0000+0 +++0+000+-0-00- +++00-00+0---0- ++000-0+000++++
++0++00+-00++0+ ++00+00+0-00+00 ++0-0-0++-00++0 ++000-0+00-0000
++0++00+-0-00-0 ++00+00+0---0-- +000+0+00-++++0 +0+++0+-00++0+0
+0000-+0000+0+0 +00++0+0-00+000 +00+0-+0-+-+-00 +0-0+0++--0++00
+0-00-++-0-+000 +000+0+00-0000- +00-0-+0+-000+- +0--0-++0--000-
+0+++0+-0000-0- +0+00-+-+000-+- +0000-+000-0-0- +00++0+0-0-0---
0+++++-000+0+0+ 0++++++-0000-0-0 0+++00-00+0000+ 0+++00-00+----0
0++000-0+0+0+++ 0++0---0++000++ 0++000-0+00-000 0+0000-+0000+0+
0+00++-+0-0-+-0 0+0-00-++-0-+00 0+00---+0+-000+ 0+0----++0--000
0+0+00-+-+-00-+ 0+0000-+00--0-0 00++++0-00+0000 00++++0-000----
00++000-0+++0++ 00++000-0+00-00 00+0000-+0+00+0 00+0--0-++00-+0
00+0++0-+-+-00- 00+0000-+00--0- 000+++00-0+++0+ 000+++00-0000-0
000+0000-+0+00+ 000+0000-+-0--0 0000++000-+0+00 0000++000-0-0--
0000000000+++++
```

In our result we have to know how the edges of our graph have been sorted. The function `snd graphP` returns the following.

Ordering of edges in the Petersen graph

```
[[0,1],[1,2],[2,3],[3,4],[4,0],
 [0,5],[1,6],[2,7],[3,8],[4,9],
 [5,7],[6,8],[7,9],[8,5],[9,6]]
```

When we use as an example the circuit in the upper right corner, we obtain the corresponding following sequence of edges of the circuit:

```
[0,1],[1,2],[2,3],[3,8],[8,5],[5,7],[7,9],[9,4],[4,0]
```

We come now to the concept of cocircuits of a graph. They play a role similar to that of circuits of a graph. A cocircuit is a minimal cut in a graph. A cut is a set of edges which causes a split of the graph into two or more connected components when we delete those edges. Minimal means that when we keep one of the edges, the split of the graph does not occur. To find the cocircuits, it is useful to have a function that calculates the number of connected components of the graph. Such a function is also useful in its own right.

We consider a list of vertices of one component of the graph and we determine all vertices that can be connected with these vertices via an edge of the graph. We repeatedly apply this function vComp, starting with a vertex of the graph and the graph as input, which leads us to all vertices of the same component of the graph. We now delete all vertices of that component in the graph and all edges that are incident with one of those vertices. We obtain the corresponding subgraph with the function vs2gr. The number of components of a graph can now be evaluated with the function nOfComp.

Number of components of a graph

```
nOfComp::GR->Int
nOfComp ([],_) = 0
nOfComp graph@(vs,es)
 |length vs== 1 = 1
 |otherwise=1+nOfComp(vs2gr(vComp[head vs]graph)graph)


vs2gr::[Int]->GR->GR
vs2gr vts (vs,es)=(vs\\ vts,[e|e<-es,e\\ vts==e])


vComp::[Int]->GR->[Int]
vComp vts (vs,es) | nvts==vts = vts
                  | otherwise = vComp nvts (vs,es)
 where
 nvts=nub(vts++concat[e\\[v]|v<-vts,e<-es,v'elem'e])
```

Let us consider how we determine the cocircuits in our Petersen graph. We first define a function es2gr that determines the subgraph of a graph in which a given set of edges has been deleted. For a given natural number *k* we consider all *k*-element subsets of 15 (the number of edges of graphP). For each case we delete corresponding edges and we calculate the remaining number of components in each case. The function `r` returns these numbers.

Subgraph after deleting edges

```
es2gr::[[Int]]->GR->GR
es2gr eds (vs,es)=(vs,es\\ eds)

r::Int->[Int]
r k= map nOfComp (map(\es->es2gr es graphP)
       [[elist!!(i-1)| i<-t]| t<-tuples k 15])
                where elist = snd graphP
```

Deleting any two edges keeps the graph connected. However, the list `res3` is not empty. We obtain

Three element edge indices of cocircuits

```
[[1, 2, 7],[1, 5, 6],[2, 3, 8],[3, 4, 9],[ 4, 5, 10],
 [6,11,14],[7,12,15],[8,11,13],[9,12,14],[10,13,15]]
```

We continue to search for 4-element subsets that do not contain any of the sublists from `res3`. This is our list `res4`. The remaining steps that we need have been written in the following. A corresponding list `res8` is equal to the empty list.

Lists of edge indices of cocircuits

```
res3=[fst el|el<-(zip (tuples 3 15) (r 3)), snd el >1]

res4=[fst el|el<-(zip (tuples 4 15) (r 4)), snd el >1,
        and[not(isSubSet e(fst el))|e<-res3]]

res5=[fst el|el<-(zip (tuples 5 15) (r 5)), snd el >1,
        and[not(isSubSet e(fst el))|e<-res3++res4]]

res6=[fst el|el<-(zip (tuples 6 15) (r 6)), snd el >1,
        and[not(isSubSet e(fst el))|e<-res3++res4++res5]]

res7=[fst el|el<-(zip (tuples 7 15) (r 7)), snd el >1,
        and[not(isSubSet e(fst el))|e<-res3++res4++res5++res6]]
```

The list `ts=res3++res4++res6++res7` of all tuples of edge indices is the source for our list of cocircuits. We can write this in a single function `res`.

However, just one function can cause problems

```
res::Int->[[Int]]
res 2 = []
res k =(res (k-1))
         ++[fst el|el<-(zip (tuples k 15) (r k)), snd el >1,
              and[not (isSubSet e (fst el))|e<-res (k-1)]]
```

The above explicit way is much faster. We use the constant function `ts` for the result of all tuples of edge indices that lead to cocircuits. The function `t2gr` generates the subgraph in which corresponding edges have been discarded.

Preparation of cocircuit result

```
ts::[[Int]]
ts=res3++res4++res5++res6++res7

t2gr::[Int]->GR->GR
t2gr t (vs,es)=(vs,[es!!(i-1)|i<-[1..length es],
                              i'notElem't       ])
ns::[Int]
ns = map nOfComp (map (\t->t2gr t graphP) ts)

cs::[[Int]]
cs = map (vComp [1]) (map (\t->delEdges t graphP) ts)
```

String of characters for cocircuit result

```
resultCocircuits::[[Char]]
resultCocircuits
 =map(\t,c)->(map(\i->(if i'notElem't
                   then '0'
                   else
                   (if head((snd graphP)!!(i-1))'elem'c
                   then '+'
                   else '-')))
              [1..15]))
  (zip ts cs)
```

We have the number of components of these subgraphs in the list `ns`. It is always 2. The function `cs` returns a list of those components that contain 1. The sign in a cocircuit tells us how the edge is directed between the two subgraphs that are split by the cocircuit. The function `resultCocircuits` finally returns the cocircuit information as a string of characters the same as in the circuit case.

A remarkable, and for the novice perhaps surprising, result will show up if the function test is applied to circuits and cocircuits. It turns out that the span of these two sets of vectors form complementary orthogonal subspaces.

Assume that all edges of the cocircuit are directed from the first component to the second. A circuit that contains such an edge has another one with opposite direction. Our product `prod` takes that into account. Circuits and cocircuits are orthogonal in this sense. We have listed a few of the 191 cocircuits via the indices of the edges and as signed vectors.

Partial cocircuit result

```
res3=[[1,2,7],[1,5,6],  ...  ,[9,12,14],[10,13,15]]
res4=[[1,2,12,15],[1,3,7,8],  ...  ,[8,10,11,15]]
res5=[[1,2,9,14,15],  ...  ,[7,9,10,13,14]]
res6=[[1,2,3,4,14,15],  ...  ,[7,8,9,10,11,14]]
res7=[[1,2,3,4,6,11,15],  ...  ,[4,5,7,8,9,11,14]]
res8=[]
```

Partial cocircuit result as character strings

```
123...           123...           123...           123...
-+0000+00000000  -000+-000000000  0+-0000-0000000  00+-0000-000000
000+-0000-00000  00000+0000-00+0  000000+0000-00+  0000000+00+0-00
00000000+00+0-0  000000000+00+0-  -+000000000+00-  -0+000++0000000
-00+0-000-00000  -000+000-00000+0 0+-0000000+0-00  0+0-000--000000
0+00-++00000000  00+-0000000+0-0  00+0-000--00000  000+-0000000+0-
00000+0+0000-+0  00000+00+0-+000  000000+0+0000-+  000000+00+0-+00
0000000+0++000-  -+000000-0000+-  -+0000000-0+-00  -0+00-00--00000
....
0+00-+00--00-+0  0+00-0+--00-+00  0+00-0+-0-000-+  0+00-00-000++--
0+00-0000-++--0  00+-0+++0000-0+  00+-0+100+-0+00  00+-0+0+0+0+00-
00+-00++0++00-0  00+0-++00--000+  00+0-+0+-0000+-  00+0-+0+0-0+-00
00+0-+0000-++0-  00+0-0++-0+-000  00+0-00+00++0--  000+-+++000-0+0
000+-++0+0-0+00  000+-+0++00+00-  000+-0+++0+00-0
```

Abstract scalar product

```
prod::[Char]->[Char]->Int
prod  []  []  = 0
prod (c:cs) (coc:cocs)
  | c=='0'||coc=='0'                              = 0+(prod cs cocs)
  |(c=='+'&&coc=='+')||(c=='-'&&coc=='-')= 1+(prod cs cocs)
  |(c=='+'&&coc=='-')||(c=='-'&&coc=='+')=-1+(prod cs cocs)

test=[prod c coc| c<-resultCircuits,  coc<-resultCocircuits]
```

We finish this interlude on graphs with a discussion of Farkas's Lemma. Oriented matroids form the most general framework in which Farkas's Lemma holds. So, it is useful to see a simple version of it. If we try to connect two vertices in a directed graph by a connected path with only positively oriented edges, we find either such a path, or a set of oriented edges that form a cut of the graph. Moreover, all the edges of the cut are directed away from the component of the graph we want to reach, compare Figure 2.12. These two cases clearly complement each other. Some Haskell code can be viewed as a constructive proof of Farkas's Lemma for graphs.

Constructive proof of Farkas's Lemma for graphs

```
lFarkas::Int->Int->GR->(Es,Bool)
lFarkas u v graph
 |u==v        = ([],True)
 |otherwise = pathCut u v [u] graph


pathCut::Int->Int->[Int]->GR->(Es,Bool)
pathCut vStart vFinal reachVs graph@(vs,es)
 |newEs == []             = ( cutEs,False)
 |vFinal 'elem' newVs = (pathEs,True )
 |otherwise = pathCut vStart vFinal (reachVs++newVs) graph
 where
 complVs=vs\\ reachVs
 cutEs =[[a,b]|[a,b]<-es, b'elem'reachVs, a'elem'complVs]
 newEs =[[a,b]|[a,b]<-es, a'elem'reachVs, b'elem'complVs]
 newVs =nub(map last newEs)
 finE  =head[[a,b]|[a,b]<-es, a'elem'reachVs, b==vFinal]
 pathEs =(fst(lFarkas vStart (head finE) graph))++[finE]
```

The decisive function has been named `pathCut`. Its input consists of the start vertex, `vStart`, the final vertex, `vFinal`, the list of vertices `reachVs` that are known to be reachable vertices and the graph. We use the complementary set of

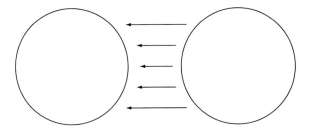

Figure 2.12 There is no oriented path from the left part to the right

vertices `complVs` to the known reachable ones. We can list those potential cut edges `cutEs` that point from the complementary set of vertices to the known reachable set of vertices and, vice versa, we can list a set of edges `newEs` that point to new reachable vertices `newVs`. We can also provide the final edge `finE` provided it does exist. When there are no new reachable vertices, that is, `newEs == [ ]`, we have found a cut. When the final vertex lies in the set of new reachable vertices, we can use recursively the function `lFarkas` to find the path from the start vertex up to the beginning of the final edge. Otherwise, we can use the function `pathCut` recursively in the case of an extended set of reachable vertices.

Examples for the Petersen graph from Page 59

```
lFarkas 2 1 graphP=([[2,3],[3,4],[4,0],[0,1]],True)
lFarkas 1 5 graphP=([[1,6],[6,8],[8,5]],True)
lFarkas 5 1 graphP=([[0,5],[1,6],[2,7],[3,8],[4,9]],False)
```

For a more detailed duality investigation within the theory of oriented matroids, see Bachem and Kern, 1992.

2.9 Summary

$$M := \begin{pmatrix} 0 & -1 & 1 \\ 0 & 0 & 1 \\ -1 & -1 & -1 \\ 0 & -1 & -1 \\ 1 & 1 & 0 \end{pmatrix}$$

	Chirotope
(1,2,3)	+1
(1,2,4)	0
(1,2,5)	-1
(1,3,4)	+1
(1,3,5)	+1
(1,4,5)	+1
(2,3,4)	+1
(2,3,5)	0
(2,4,5)	+1
(3,4,5)	-1

Chirotope

```
[ ([1],[[+2,-4],[-5],[+3]]),
  ([2],[[+1,+4],[-3,+5]  ]),
  ([3],[[+1],[+2,-5],[-4]]),
  ([4],[[+1,+2],[-5],[+3]]),
  ([5],[[+1],[+4],[-2,+3]])]
```

Hyperline Sequences

```
          1 2 3 4 5
(0,0,-,0,+)
(0,0,+,0,-)
(0,+,0,-,-)
(0,-,0,+,+)
(0,+,-,-,0)
(0,-,+,+,0)
(+,0,0,+,0)
(-,0,0,-,0)
(+,+,0,0,-)
(-,-,0,0,+)
(+,+,-,0,0)
(-,-,+,0,0)
```

Cocircuits

(1,2)	-1
(1,3)	-1
(1,4)	0
(1,5)	-1
(2,3)	+1
(2,4)	-1
(2,5)	+1
(3,4)	-1
(3,5)	0
(4,5)	+1

Dual Chirotope

$$(\{1,-4\},\{-2\},\{-3,-5\})$$

Dual Hyperline Sequences

```
          1 2 3 4 5
(+,-,0,-,0)
(-,+,0,+,0)
(0,+,+,0,+)
(0,-,-,0,-)
(+,0,+,-,+)
(-,0,-,+,-)
```

Circuits

Dual Cocircuits

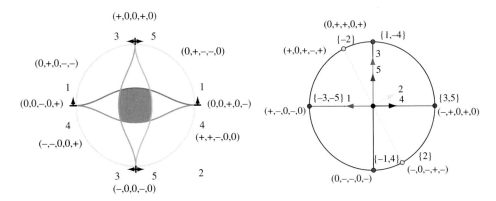

Figure 2.13   Primal and dual sphere system with covectors and vectors

We summarize data structures containing the oriented matroid information of our given matrix M. The multitude of ways to look at a matrix should no longer lead to confusion.

It should be easy to understand the set of vectors and the set of covectors for our example.

We have seen that we can obtain different combinatorial data from a matrix: for example, the chirotope, or its dual chirotope; the set of hyperline sequences, or the corresponding dual set of hyperline sequences; all cocircuits, or all dual cocircuits (= circuits) that are given by the sphere system, or by the dual sphere system, respectively.

It turns out that for a given matrix, we can switch from one corresponding data structure to another and vice versa. The combinatorial information is the same in all cases. If we characterize this combinatorial information by axioms, we have to generalize the data structure that we have derived from matrices, which leads us to *oriented matroids* in the general case. For the transition from a chirotope to its dual, we have presented a Haskell program. The transition from a chirotope to its sphere system can be obtained via the omawin program. The proof of equivalence in the rank 3 case, a topological representation theorem for oriented matroids, will be given in the next chapter.

Characterizations of oriented matroids have been studied in many ways with various motivations. As a consequence, in the theory of oriented matroids, we are faced with terminology that comes from various roots, in fields that are usually not very closely related.

- F. Levi was the first, see Levi, 1926, to study oriented matroids in rank 3 as generalized topological lines in the plane that he called *pseudolines*.
- Sphere systems, generalization of pseudoline arrangements to arbitrary dimensions, turn out to be oriented matroids. We have early contributions, see Folkman and Lawrence, 1978, and Edmonds and Mandel, 1982.

- Chirotopes, first independently studied in Gutierrez Novoa, 1965, and later in Fenton, 1982, and in Dreiding, Dress, and Haegi, 1982. Bokowski's approach that started in 1979 and was first applied implicitly in Altshuler, Bokowski, and Steinberg, 1980 has led to oriented matroids. Bokowski required essentially that the signed bases do not violate GP-relations directly.

- Sets of hyperline sequences have been studied by Bokowski since 1979 but also studied independently by I. Streinu in the context of computational geometry when using ideas of E. Goodman and R. Pollack, that is, *order types* and *allowable sequences*. They are characterized essentially by requiring that the signs of bases (signs of abstract determinants) are consistent in all hyperlines, Bokowski, King, Mock, and Streinu, 2005.

- That concept also leads immediately to covectors. But they can be characterized directly as well.

- Via circuit axioms Las Vergnas introduced oriented matroids by generalizing oriented graphs, see Las Vergnas, 1975.

- Generalizing the dual concept, known from linear programming, from vector spaces to the most general useful case leads again to oriented matroids (Bland's approach).

- We can say that oriented matroids differ from matroids by just the underlying field (GF_3 instead of GF_2) if we use an algebraic variety approach, see Bokowski, Guedes de Oliveira, and Richter-Gebert, 1991.

- Oriented matroids have been characterized by order functions, independently, in Kalhoff, 1996 and Jaritz, 1996.

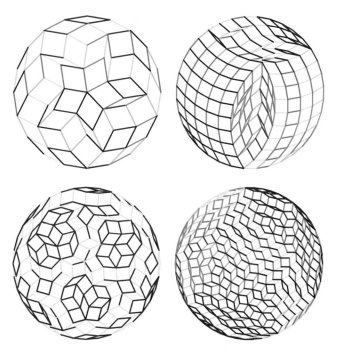

Figure 2.14 Zonotopal tilings corresponding to visable traces in Figure 1.16

- Knuth recommended in 1992, see Knuth, 1992, the study of certain hull systems which turned out to be oriented matroids.
- Klin and coauthors studied oriented matroids when they investigated questions in chemistry, see Klin, Tratch, and Treskov, 1989.
- Oriented matroids of rank 3 are in a sense generalized Platonic Solids; a corresponding definition in Bokowski, Roudneff, and Strempel, 1997 uses Petrie polygons, a notion introduced by Coxeter.

We finally mention that oriented matroids in rank 3 can be characterized by zonotopal tilings. Figure 2.14 shows examples of what these objects look like, see Bohne, 1992, and Richter-Gebert and Ziegler, 1994.

2.10   What is the oriented matroid of a matrix?

We assume that the reader is now ready to fully understand a page from Ziegler's paper entitled *Oriented Matroids Today*, see Ziegler, 1996, where he tries to answer the question: *what is an oriented matroid?* Of course, he gives his answer first for a finite, spanning, sequence of vectors only, that is, for a given full rank matrix like our example matrix in this chapter. The transition to the more general concept of an oriented matroid has still to be worked out. This will be done in the rank 3 case in our next chapter. We now copy two passages from Ziegler's paper: first a general remark about oriented matroids, finally a list of five concepts that store the information of the oriented matroid of a matrix.

Oriented matroids are both important and interesting objects of study in Combinatorial Geometry and indispensable tools of increasing importance and applicability for many other parts of Mathematics. The main parts of the theory and some applications were, in 1993, compiled in the comprehensive monograph Björner *et al.*, 1993.

For other introductions and surveys, see Bachem and Kern, 1992, Bokowski and Sturmfels, 1989b, Bokowski, 1993, Goodman and Pollack, 1993, Ziegler, 1994, Chapters 6 and 7, and Richter-Gebert and Ziegler, 1997.

Let $V = (v_1, v_2, \ldots, v_n)$ be a finite, spanning, sequence of vectors in $\mathbb{R}^r$, that is, a finite *vector configuration*. With this vector configuration, one can associate the following sets of data, each of them encoding the *combinatorial structure* of V.

- The *chirotope* of V is the map

$$\chi_v : \{1, 2, \ldots, n\}^r \longrightarrow \{+, -, 0\}$$
$$(i_1, i_2, \ldots, i_r) \longmapsto \text{sign}(\det(v_{i_1}, v_{i_2}, \ldots, v_{i_r}))$$

that records for each r-tuple of the vectors whether it forms a positively oriented basis of $\mathbb{R}^r$, a basis with negative orientation, or not a basis.

- The set of *covectors* of V is

$$\mathcal{V}^*(V) := \left\{ \left(\operatorname{sign}(a^t v_1), \ldots, \operatorname{sign}(a^t v_n)\right) \in \{+, -, 0\}^n : a \in \mathbb{R}^r \right\},$$

that is, the set of all partitions of V (into three parts) induced by hyperplanes through the origin.

- The collection of *cocircuits* of V is the set

$$\mathcal{C}^*(V) := \left\{ \left(\operatorname{sign}(a^t v_1), \ldots, \operatorname{sign}(a^t v_n)\right) \in \{+, -, 0\}^n : a \in \mathbb{R}^r \right.$$

$$\left. \text{is orthogonal to a hyperplane spanned by vectors in } V \right\}$$

of all partitions by "special" hyperplanes that are spanned by vectors of the configuration V.

- The set of *vectors* of V is

$$\mathcal{V}(V) := \left\{ \left(\operatorname{sign}(\lambda_1), \ldots, \operatorname{sign}(\lambda_n)\right) \in \{+, -, 0\}^n : \lambda_1 v_1 + \ldots + \lambda_n v_n = 0 \right.$$

$$\left. \text{is a linear dependence between vectors in } V \right\}.$$

- The set of *circuits* is

$$\mathcal{C}(V) := \left\{ \left(\operatorname{sign}(\lambda_1), \ldots, \operatorname{sign}(\lambda_n)\right) \in \{+, -, 0\}^n : \lambda_1 v_1 + \ldots + \lambda_n v_n = 0 \right.$$

$$\left. \text{is a } \textit{minimal} \text{ linear dependence between vectors in } V \right\}.$$

3

From matrices to rank 3 oriented matroids

In this chapter we begin a transition from the combinatorial matrix concepts of the last chapter to its oriented matroid extension. Our goal is a gentle introduction of the general concept in the rank 3 case. We start with some general code that determines the chirotope of a matrix even in the arbitrary rank case. We discuss in detail the rank 2 case of oriented matroids in which the matrix cases form the most general examples. The transition from chirotopes to hyperlines and vice versa is a good preparation for the rank 3 case in which the matrices no longer provide the most general oriented matroids. A planar point set example will be used next to work in the rank 3 case. We compare the corresponding rank 2 pictures and its hyperline data structure, the rank 2 contractions of the oriented matroid of rank 3.

The following first definition of a uniform chirotope in rank 3 requires an abstract sign of determinant function to exist. Next we discuss how the convex hull structure can be seen in an oriented matroid of rank 3. A fundamental concept is to form equivalence classes within the set of oriented matroids. Reorienting an oriented matroid, relabeling the elements, and/or a sign reversal of all signed bases are in many cases not essential changes. We discuss these concepts and show that in the uniform case in rank 3 there is up to these changes precisely one oriented matroid with five elements. From all possible axioms for oriented matroids in rank 3, we introduce the topological sphere system concept as pseudoline arrangements, the chirotope concept that uses (ordinary) matroids and Grassmann–Plücker relations, and hyperline sequences that have advantages for algorithmic applications. We show that these three axiom systems are equivalent and we provide a few oriented matroids that are not representable as matrices.

For a survey article of rank 3 oriented matroids see Goodman 1997.

3.1   From a matrix to its chirotope

Computing the chirotope, one possible data structure of an oriented matroid, from a given matrix is a simple matter. We next provide some Haskell code for that. The given matrix is seen as a constant matrix function m.

We use our (5×3)-matrix M from Chapter 1

```
m::MA  -- type MA = [[Integer]]
m=[[ 0,-1, 1],
   [ 0, 0, 1],
   [-1,-1,-1],
   [ 0,-1,-1],
   [ 1, 1, 0]]
```

For the types that we use for our functions see Page 303. The following Haskell program assumes a matrix as input and returns the corresponding chirotope. We have implemented the Laplacian rule for evaluating the determinant knowing that there are better ways in higher dimensions.

Determinant of a matrix

```
det::MA -> Integer
det m
 |n == 1  = head (head m)
 |otherwise=sum(map
  (\i->((-1)^i))*(head(m!!i))*(det
  [(map tail m)!!l|l<-[0..n-1],l/=i]))[0..n-1])
  where n = length m
```

The function `det` returns the value of the determinant of the input matrix. Of course, we assume that the input is a square matrix. The case of a (1×1)-matrix is clear. For implementing the Laplacian rule, we have used from the Haskell prelude file the functions head, map, sum, (!!), tail, (*), compare the Haskell primer, see Appendix A.

Determinants of submatrices

```
dets::[[Int]]->MA-> [Integer]
dets sets matrix = [det[matrix!!(i-1)|i<-set]|set<-sets]
```

With the function `dets` we get all determinants of submatrices with row indices in the given list of tuples. We have used the map function that evaluates, in this case, the sign of the determinant for each entry in the list.

How to obtain the chirotope of a matrix

```
m2Chi::MA->[OB]
m2Chi m=zip trn (map toInt (map signum(dets trn m)))
        where n=length m
                r=length(head m)
                trn=tuples r n
```

For our example matrix we use `trn = tuples 3 5`; we apply the function `signum` from the prelude module of Haskell and end up with the following.

The chirotope of our sample matrix M

```
m2Chi m
[([1,2,3], 1),
 ([1,2,4], 0),
 ([1,2,5],-1),
 ([1,3,4], 1),
 ([1,3,5], 1),
 ([1,4,5], 1),
 ([2,3,4], 1),
 ([2,3,5], 0),
 ([2,4,5], 1),
 ([3,4,5],-1)]
```

Although our intention in this chapter is to emphasize the rank 3 case, we see in this case that the general rank case requires essentially the same work. Our functions can be applied to arbitrary matrices with r columns and n rows and $n \geq r$.

3.2   The basic rank 2 case

We come back to our main objective. We aim at a further understanding of the oriented matroid data structure that arises from the matrix concept. A fundamental understanding comes from a detailed study of the rank 3 case. However, we first start in the rank 2 case.

Let us consider a matrix with n rows of non-zero vectors with two coordinates each. We divide each vector by its Euclidean norm. Thus we have n points on the unit circle. We use the labels $\{1, \ldots, n\}$ of the rows as the ordering of the vectors.

The orientation for any ordered pair of labels is given by the sign of the determinant of the corresponding (2×2)-submatrix. We can have, for example,

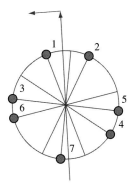

Figure 3.1   A rank 2 example

equal rows or rows that differ just by multiplying each component by −1. The
oriented matroid information of the matrix is just the ordering of the labels on
one half of the circle, say beginning with the smallest label and going counter-
clockwise around the circle. We write all labels with equal row vectors as a set
of these labels together with the negative of all labels of those rows that differ
just by multiplying them with −1. The example in Figure 3.1 can be assumed
to represent an $(n \times 2)$-matrix in which each row vector has been normalized.
When we write this example as some Haskell code, we have forgotten metrical
distances on the boundary of the circle. We have written its normalized hyperline
sequence.

Normalized hyperline sequence for Figure 3.1.

```
[[1],[-4],[3,-5],[6],[-2],[7]]
```

In this rank 2 case, oriented matroids are just equivalence classes of matrices
as described above, that is, a list of lists (the first one containing 1) such that all
elements appear precisely once in one of these lists with either a positive or a
negative sign. If we start with such a list of lists (an oriented matroid of rank 2),
we can easily find a corresponding matrix that leads to it as described above. We
can just write down unit vectors with increasing angles with respect to the first
vector as long as the angles lie between zero and 180°. A negative sign in the
sequence gives rise to the opposite direction of the vector. Our remark that we
can always find a vector representation of such a normalized hyperline sequence,
will no longer hold true in the general case of oriented matroids of higher rank.
This fact only holds in the rank 2 case.

Let us determine the chirotope information from the above normalized hyperline
sequence. We obtain the following as a result of the function `tuples 2 7`.

All pairs for which we have to determine signs

```
[[1,2],[1,3],[1,4],[1,5],[1,6],[1,7],[2,3],
 [2,4],[2,5],[2,6],[2,7],[3,4],[3,5],[3,6],
 [3,7],[4,5],[4,6],[4,7],[5,6],[5,7],[6,7]]
```

We can determine the chirotope from the given hyperline with the following Haskell function.

Chirotope of rank 2 from a given hyperline

```
hyp2ChiR2::OM2->[OB]
hyp2ChiR2 [] = []
hyp2ChiR2 (h:rest)
 = [norm([h!!(i-1),h!!(j-1)],0)|[i,j]<-tuples 2 l]
   ++[norm([h!!i,(rest!!j)!!k],1)
     | i<-[0..l-1], j<-[0..lr-1],
       k<-[0..length(rest!!j)-1]]
 ++hyp2ChiR2 rest    where l = length h
                           lr= length rest
```

The function `hyp2ChiR2` takes one list `h` of `OM2` at a time. Pairs of indices in `h` have zero orientation. We compare each single element in `h` with other indices in the sequence and return the normalized form of the ordered pair. Finally, we apply `hyp2ChiR2` on the remaining part `rest` and we sort the oriented pairs of indices to obtain the following.

Resulting chirotope

```
[(([1,2],-1), ([1,3], 1), ([1,4],-1),
  ([1,5],-1), ([1,6], 1), ([1,7], 1),
  ([2,3], 1), ([2,4],-1), ([2,5],-1),
  ([2,6], 1), ([2,7],-1), ([3,4], 1),
  ([3,5], 0), ([3,6], 1), ([3,7], 1),
  ([4,5], 1), ([4,6],-1), ([4,7],-1),
  ([5,6],-1), ([5,7],-1), ([6,7], 1)]
```

The reverse function `chi2HypR2` converting a chirotope of rank 2 into its corresponding normalized hyperline information is a little bit more complicated. We are going to study this now.

From the function `r2ex` we get our example. The function `chi2HypR2` just does a final sorting according to the function `chi2ordf`. We should assume that all lists in the sequence have been prepared earlier and the sorting is done last. So, we look at the function `preHypR2` that ensures this. We have to understand the function `signIns` that inserts the right signs into the lists in the sequence and the function `genSets` that generates these sets without signs. These two steps can be studied independently and it should be clear what these functions do from a careful analysis of the functions.

A chirotope example in rank 2 with eight elements

```
r2ex::[OB]
r2ex
 = [([1,3], 0),([1,4], 1),([1,5],-1),([1,6], 1),
     ([1,7], 1),([1,8], 1),([1,9], 1),([3,4], 1),
     ([3,5],-1),([3,6], 1),([3,7], 1),([3,8], 1),
     ([3,9], 1),([4,5], 1),([4,6], 0),([4,7], 1),
     ([4,8], 1),([4,9], 1),([5,6],-1),([5,7],-1),
     ([5,8],-1),([5,9],-1),([6,7], 1),([6,8], 1),
     ([6,9], 1),([7,8], 1),([7,9], 1),([8,9],-1)]
```

Generation of unordered sets of hyperline

```
genSets::[OB]->[[Int]]
genSets chi = sort(glue chi (tuplesL 1 els))
 where els=sort (nub (concat (map fst chi)))

glue::[OB]->[[Int]]->[[Int]]
glue [] pre = pre
glue (([u,v],s):xs) pre
  |s==0= glue xs([set|set<-pre,intersect set[u,v]==[]]
        ++[nub(concat[a++b|a<-pre,b<-pre,
                            u'elem'a, v'elem'b])])
  |otherwise = glue xs pre
```

When we use the function `genSets` and we apply it to `r2ex` we obtain the matroid information, that is, sets of indices that belong to the same set have been found.

genSets r2ex = [[1,3],[4,6],[5],[7],[8],[9]]

At the beginning all (positive) elements `els` of the chirotope form single element sets and they are sorted. The function `glue` forms the union of such sets when a corresponding zero orientation of the oriented matroid requires this. Observe that the minimal elements from each sublist still form an ordering on these sublists. In particular, we have that the first sublist contains the minimal element.

The next step is to insert the correct signs for all elements. We first observe that the chirotope of rank 2 induces an ordering of pairs of signed indices of our elements. This is clear from the previous wheel picture of Figure 3.1. A corresponding Haskell program is the following. The image of our function `chi2ordf` is indeed again a function `f:[Int]->[Int]->Ordering`.

Ordering function of a chirotope of rank 2

```
chi2ordf::[OB]->([Int]->[Int]->Ordering)
chi2ordf chi = f
  where f::[Int]->[Int]->Ordering
        f a b|norm([head a,head b],1)'elem'chi=LT
             |otherwise                        =GT
```

Now we look at our list of sublists

$$[[1,3],[4,6],[5],[7],[8],[9]]$$

and we compare the minimal element in the first sublist with all other elements from the remaining sublists. The ordering function of our chirotope induces the proper signs for all those elements.

With `signsTail (chi2ordf r2ex) (genSets r2ex)` we obtain:

$$[[1,3],[4,6],[-5],[7],[8],[9]].$$

Inserting the sign in the tail of sublists

```
signsTail::OF->[[Int]]->[[Int]]
signsTail ord (l:ls)=[l]++[map signT set|set<-ls]

  where
  signT::Int->Int
  signT el |ord [min] [el] ==LT =  el
           |otherwise            = -el
```

We can now insert the proper signs for the remaining elements in the first sublist. We compare these elements with the last element in the last sublist. Again the ordering function induces the proper signs for all those elements.

Inserting the sign in the head of sublists

```
signsHead::OF->[[Int]]->[[Int]]
signsHead ord (l:ls)
 =[[min]++map(\el->signH ord el) (l\\[min]) ]++ls
 where min = minimum l
        signH::OF->Int->Int
        signH ord el
          |ord [el] [last(last ls)]]==LT =  el
          |otherwise                      = -el
```

The application of the function `signsHead` can be written in a better way as in the function `siIn`. The function `siIn r2ex` leaves the former list of sublists unchanged.

Complete sign insertion

```
siIn::[OB]->[[Int]]
siIn chi=signsHead ord (signsTail ord (genSets chi))
      where ord = chi2ordf chi
```

What we finally have to do is a sorting of all sublists according to our ordering function of the chirotope. We call the corresponding final function `chi2HypR2`.

Generation of hyperline, final step

```
chi2HypR2::[OB]->OM2
chi2HypR2 chi = sortBy (chi2ordf chi) (siIn chi)
```

With `chi2HypR2 r2ex` we arrive at

$$[[1,3],[-5],[4,6],[7],[9],[8]]$$

Now we have studied both directions in the rank 2 case, the transition from the chirotope to its hyperline sequence and the transition from the hyperline sequence to its chirotope. For later applications, it is good to have a sound understanding of how to switch back and forth between these two models. However, as long as we have this easy rank 2 case, there is nothing that goes beyond the well-known matrix cases.

3.3 A planar point set example

In this section we discuss a Haskell program that computes the hyperline sequences from a point configuration in the plane. This should familiarize the reader once more with the concept of hyperlines and it should emphasize the advantages of this data structure.

Moreover, by analysing the transition from the point set to its hyperline information, we see that the chirotope information of the matrix can be used as an intermediate data structure. In other words, we understand how to convert a rank 3 chirotope of a point set to its hyperline sequences. On the other hand, we will see later that the chirotope can be computed from hyperline sequences.

The theory of oriented matroids deals with an extension of the former data structures, hyperline sequences, and chirotopes, from the matrix case to a more general case for which the former transitions still make sense. We will deal with a certain combinatorially closed system of axioms that includes all matrix cases.

Like in the matroid case, where we have the famous title of an article attributed to Vamos: *the missing axiom is lost forever*, the attempt to characterize precisely those chirotopes, hyperline sequences, or other data sets of general oriented matroids comming from matrices, is doomed to failure, see Bokowski and Sturmfels, 1989a, for an infinite family of minor-minimal nonrealizable 3-chirotopes.

For the nine point planar example in Figure 3.2 we determine its normalized hyperline sequence. The function `mex` contains the affine point set information as a matrix.

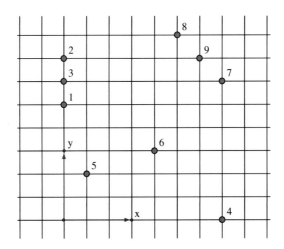

Figure 3.2   A planar example

Given affine coordinates of the example

```
mex::MA
mex=[[0,5],[0,7],[0,6],[7,0],[1,2],[4,3],[7,6],[4,9],[6,7]]
```

As an intermediate step we use the chirotope of the matrix of homogeneous coordinates `homCoord`. We generate homogeneous coordinates by adding a first column with entries 1 by `homCoord`.

Homogeneous coordinates

```
homCoord::MA
homCoord = map ([1]++) mex
```

We evaluate the corresponding chirotope of the matrix via the function `m2Chi` on Page 74.

From this chirotope we generate all contractions at single elements. The meaning of a contraction at a point is that we look at the circular sequence in which the remaining points are seen from this point, thereby forgetting this central point. The former function `chi2HypR2` then leads to the corresponding hyperline sequence of our example when we use the function `hypex`.

Hyperline sequence of example

```
hypex::[(([Int],OM2)]
hypex=map(\i->([i],chi2HypR2(ctrElChi i(m2Chi mex))))[1..9]
```

Hyperline sequences of the given example

```
[([1], [[2,3],[-5],[-4],[-6],[-7],[-9],[-8]]),
 ([2], [[1,3],[5],[4,6],[7],[9],[8]]),
 ([3], [[1,-2],[5],[4],[6],[7],[9],[8]]),
 ([4], [[1],[5],[-7],[-9],[-8],[-2,-6],[-3]]),
 ([5], [[1],[-4],[-6],[-7],[-9],[-8],[-2],[-3]]),
 ([6], [[1],[5],[-7],[-9],[-8],[4,-2],[-3]]),
 ([7], [[1],[5],[6],[4],[-8,-9],[-2],[-3]]),
 ([8], [[1],[5],[6],[4],[7,9],[-2],[-3]]),
 ([9], [[1],[5],[6],[4],[7,-8],[-2],[-3]])]
```

Compare the result with all corresponding rank 2 pictures in Figure 3.3 to see that these concepts do not really differ.

Independently of this example we can write a general function `chi2HypR3` that starts with a rank 3 chirotope and returns the hyperline data structure in rank 3.

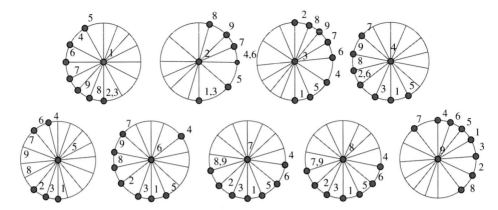

Figure 3.3   All corresponding rank 2 pictures

Generation of hyperlines in rank 3

```
chi2HypR3::[OB]->[([Int],OM2)]
chi2HypR3 chi=map(ı->([i],chi2HypR2(ctrElChi i chi)))els
        where els=nub(concat(map fst chi))
```

Chirotope of hyperline sequences in rank 3

```
hyp2ChiR3::[([Int],OM2)]->[OB]
hyp2ChiR3 [] = []
hyp2ChiR3 (h:rest)
 =nub(sort([norm([a,b,c],1)|
            a<-fst h,[i,j]<-tuples 2 l,
            b<-((snd h)!!(i-1)),
            c<-((snd h)!!(j-1)) ]
   ++[norm(pair++[c],0)|pair<-(tuplesL 2(fst h)),
                        c  <-concat(snd h)]
   ++[norm([a]++pair,0)|a<-fst h,i<-[1..l],
                        pair<-tuplesL 2((snd h)!!(i-1))]
   ++ hyp2ChiR3 rest))
                        where l = length (snd h)
```

The reverse problem of finding the chirotope from hyperline sequences was easy in the rank 2 case. The same holds true in the rank 3 case. We present a corresponding function in the rank 3 case, `hyp2ChiR3`, for calculating the chirotope if the hyperline sequences are given.

The function `hyp2ChiR3` that finds the chirotope from given hyperline sequences in rank 3 is very similar to the rank 2 case. The non-zero values of ordered triples occur if we pick one element in the index set of the hyperline and two elements from different lists in the sequence.

For later considerations we include the non-uniform case as well. Zero orientations of triples occur either if two different elements of a triple occur in the same index set or when they belong to the same set in a sequence. When we apply `hyp2ChiR3` to the hyperline sequences of the previous example, we obtain again the original chirotope.

3.4   Uniform chirotopes and hyperline sequences

Observe that the orientation of a triangle appears three times in the hyperline sequences. These orientations are of course equal since the original picture was a point configuration in the plane.

The requirement that orientations of triangles coming from different lines of hyperline sequences should not contradict each other gives rise to a definition of rank 3 uniform oriented matroids. A list of non-zero signs for all ordered triples out of n elements forms an abstract *uniform chirotope in rank 3* if and only if the corresponding sequences that we obtain via the function `chi2HypR3` provides equal signs for all oriented triples, independently of the line within the list of sequences. The image of the function `chi2HypR3` will be called abstract *uniform hyperline sequences in rank 3*. Thus we have reached the extension of chirotopes or hyperline sequences of matrices (two versions of the oriented matroid information of a matrix) to the general case of an (abstract) uniform chirotope or (abstract) uniform hyperline sequences of rank 3.

We conclude: we have seen in our Haskell program that we can start from the chirotope and we get the hyperline sequences via the function `chi2HypR3`. On the other hand, we can obtain, from the hyperlines in rank 3, the corresponding chirotope. This function `hyp2ChiR3` and its inverse `chi2HypR3` do not depend on the original matrix. We can argue on a purely abstract level. In other words, we have reached the oriented matroid reasoning in rank 3. This already forms an equivalence of chirotopes and hyperline sequences in the uniform rank 3 case and the Haskell program tells us how to convert both data structures.

3.5   Convex hull of a chirotope in rank 3

Which elements are vertices of the convex hull of all points is clear from the corresponding nine rank 2 pictures in Figure 3.3. In order to check whether an

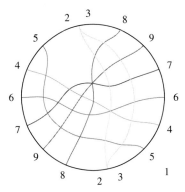

Figure 3.4   Unoriented pseudoline arrangement of Figure 3.2

element is a vertex, you look at the circle in which that element is the centre. The element is a vertex if and only if it lies outside of the convex hull formed by the remaining elements. In our example we see that the elements 1,2,4,5,7,8 are vertices. Seeing the vertices in Figure 3.2 cannot be considered to be a practical algorithm for finding them when the matrix is given. Moreover, we see that the vertices of the convex hull can be determined from the hyperlines or from the chirotope even when there might not exist a corresponding matrix.

From Chapters 1 and 2 we have some experience of changing from one oriented matroid model to another. Let us look at the convex hull structure in another model. We have depicted in Figure 3.4 the unoriented version of the topological representation for the above chirotope. From the chirotope information we obtain this picture via the omawin program. The lines in the given example containing the points (1,2,3), (2,4,6), and (7,8,9) correspond to the intersections of three pseudolines. The labels of the vertices of the convex hull correspond to labels of pseudolines that bound a two-dimensional cell in Figure 3.4, the unique topological hexagon.

We can of course also determine easily the sequence of vertices that occur when we go around the convex hull. This information can be seen directly in Figure 3.4. However, the adjacent vertices can also be read off from Figure 3.3 or from the hyperlines. We come back to determine the convex hull when we study point sets in arbirary dimensions.

3.6   Reorientation, relabeling, sign reversal

Already, in the rank 2 case, where oriented matroids always have a vector model, we observe that we can reorientate an element. Reorientating an element means that we replace the element with its negative. We can also reorientate a set of elements by reorientating all the elements of the set. There are many properties of oriented matroids that are invariant under reorientating the elements of a set. Therefore, we introduce the equivalence class of oriented matroids with respect to reorientations and call it the *reorientation class*. This concept can be defined for oriented matroids in higher rank in the same way. By reorientating an element in our above planar example, we should think of the vector model in $\mathbb{R}^3$, where each point in the given plane (not passing through the origin) will be interpreted as a non-zero vector in $\mathbb{R}^3$. Reorientating an element (vector) means replacing it with its negative.

There are also many properties that are invariant under a permutation of all the elements of a set. In those cases we use the concept of the *equivalence class of oriented matroids with respect to reorientations and relabelings*.

For matrices the equivalence class under relabelings means that we are interested in properties that do not change when we permute the rows of the matrix. The equivalence class under reorientations means that we are not interested in the sign of the direction in which the row vectors point. In the central plane model, we use unoriented central planes.

When we reverse the sign of all determinants of a chirotope simultaneously, this is just a change of the orientation of the underlying space. The geometric properties that we study are invariant under this operation that we call *sign reversal*, compare also with Page 50.

We have seen a first definition of a uniform chirotope on Page 83. Orientations should not lead to contradictions, that is, we can determine hyperline sequences such that the three ways of extracting on orientation for an oriented triangle should be consistent.

For a later decisive application within an equivalence proof of different models for an oriented matroid, we try to determine all different uniform chirotopes of rank 3 with five elements up to sign reversal, up to relabelings, and up to sign reversals. The corresponding implementations are useful in their own right. Implementing the sign reversal is perhaps the easiest task that we can do for chirotopes. We do this immediately for the arbitrary rank case with the function `signRevChi`.

Sign reversal of a chirotope

```
signRevChi::[OB]->[OB]
signRevChi chi = map (\(tu,s)->(tu, negate s)) chi
```

When we have to reorientate a single tuple at an element, the corresponding function `reorChiTuple` does this for us. When the element is contained in the tuple, the sign has to change, otherwise the sign remains. A repeated application for all tuples does the function `reorElChi`. And when we have to reorientate several elements of a list, we use the corresponding function `reorSetChi`.

Reorientating sets of elements

```
reorSetChi::[Int]->[OB]->[OB]
reorSetChi [] chi = chi
reorSetChi (el:els) chi =reorSetChi els (reorElChi el chi)


reorElChi::Int->[OB]->[OB]
reorElChi el chi = map (reorChiTuple el) chi


reorChiTuple::Int->OB->OB
reorChiTuple x (tu,sign) | x'elem'tu = (tu,-sign)
                         | otherwise = (tu, sign)
```

The function `reorLists` provides us with all possible different lists at which we can reorientate a given chirotope with n elements.

List of all sublists

```
reorLists::Int->[[Int]]
reorLists n = concat(map (\i-> tuples i n) [0..n])
```

The function `reorLists 5` returns the following.

List of all sublists for five elements

```
[[],
 [1],[2],[3],[4],[5],
 [1,2],[1,3],[1,4],[1,5],[2,3],
 [2,4],[2,5],[3,4],[3,5],[4,5],
 [1,2,3],[1,2,4],[1,2,5],[1,3,4],[1,3,5],
 [1,4,5],[2,3,4],[2,3,5],[2,4,5],[3,4,5],
 [1,2,3,4],[1,2,3,5],[1,2,4,5],[1,3,4,5],[2,3,4,5],
 [1,2,3,4,5]]
```

The following function `permsSet` provides us with the list of all permutations of a given list.

All permutations of a given list

```
permsSet::[Int]->[[Int]]
permsSet l
  |length l == 1 = [l]
  |otherwise=concat(map(\e->map([e]++)(permsSet(l\\[e])))l)
```

The function `permsSet` returns for the input `[1,2,4]` the list of sublists `[[1,2,4],[1,4,2],[2,1,4],[2,4,1],[4,1,2],[4,2,1]]`. We obtain from such a single sublist the permutation as a function when we use the function `perm`.

Permutations as functions

```
perm::[Int]->(Int->Int)
perm list = (\i-> list!!(i-1))
```

Such a function `perm` is used as input `pi` for the function `relabelST` to permute the elements of a tuple. Finally we change the signed tuple according to the function `norm` from Page 21. This is done for all signed tuples of a given chirotope with the function `relabelChi`.

Relabeling of a chirotope

```
relabelChi::(Int->Int)->[OB]->[OB]
relabelChi pi (st:sts)
  = sort ([relabelST pi st]++(relabelChi pi sts))

relabelST::(Int->Int)->OB->OB
relabelST pi (tu,s) = norm (map pi tu,s)
```

So far we have defined the functions for the chirotope model of oriented matroids. Relabeling, reorientating, and sign reversal are basic operations in many applications. It makes sense to write corresponding functions for the hyperline sequences model as well. We will do this in the following.

We come back to our aim to show that there is only one uniform 5-element chirotope or uniform hyperline configuration in rank 3 up to relabeling, reorientation, and sign reversal.

We start with a uniform hyperline sequences representation for 5 elements in rank 3. Without loss of generality, we can assume the first row of our hyperline representation is of the form $([1], [[2], [3], [4], [5]])$. Otherwise, we relabel the elements and we get the proper signs by a possible reorientation. Hyperline sequences have a unique sign vector that we have as a chirotope. The first hyperline sequence determines six such signs of this sign vector of the chirotope. We consider now all 16 possible extensions of sign vectors according to the function `all5`.

All candidates of chirotopes

```
all5::[[Or]]
all5=map([1,1,1,1,1,1]++)[[a,b,c,d]|a<-s,b<-s,c<-s,d<-s]
    where s = [-1,1]
```

For the function `chi2HypR3` from Page 82, the sign vector need not be a chirotope. The function `cand` provides 16 potential hyperline sequences.

All candidates of hyperline sequences

```
cand::[[([Int],OM2)]]
cand = map chi2HypR3 (map (zip (tuples 3 5)) all5)
```

If the extracted signs from our hyperline sequences are not consistent, we discard the candidate because of the definition of hyperline sequences in the uniform case. The function `hyp2ChiR3` would lead to more signed simplices than the chirotope should have. We are left with only eight examples of remaining hyperline sequences after using the function `rest`.

Remaining hyperline sequences

```
rest::[[([Int],OM2)]]
rest=[hyp|hyp<-cand,length(nub(sort(hyp2ChiR3 hyp)))==10]
```

We obtain the following eight hyperline sequences. We have written the whole data set. Seeing our rather short Haskell functions might have hidden that our data structures are in general difficult to handle by hand.

```
[(([1],[[2],[ 3],[ 4],[ 5]]),    [(([1],[[2],[ 3],[ 4],[ 5]]),
 ([2],[[1],[-5],[-4],[-3]]),      ([2],[[1],[-5],[-4],[-3]]),
 ([3],[[1],[-5],[-4],[ 2]]),      ([3],[[1],[-4],[-5],[ 2]]),
 ([4],[[1],[-5],[ 3],[ 2]]),      ([4],[[1],[ 3],[-5],[ 2]]),
 ([5],[[1],[ 4],[ 3],[ 2]]))]     ([5],[[1],[ 3],[ 4],[ 2]]))]
```

```
[([1],[[2],[ 3],[ 4],[ 5]]),        [([1],[[2],[ 3],[ 4],[ 5]]),
 ([2],[[1],[-4],[-5],[-3]]),         ([2],[[1],[-4],[-3],[-5]]),
 ([3],[[1],[-4],[-5],[ 2]]),         ([3],[[1],[-4],[ 2],[-5]]),
 ([4],[[1],[ 3],[ 2],[-5]]),         ([4],[[1],[ 3],[ 2],[-5]]),
 ([5],[[1],[ 3],[ 2],[ 4]])]         ([5],[[1],[ 2],[ 3],[ 4]])]
```

```
[([1],[[2],[ 3],[ 4],[ 5]]),        [([1],[[2],[ 3],[ 4],[ 5]]),
 ([2],[[1],[-5],[-3],[-4]]),         ([2],[[1],[-3],[-5],[-4]]),
 ([3],[[1],[-5],[ 2],[-4]]),         ([3],[[1],[ 2],[-5],[-4]]),
 ([4],[[1],[-5],[ 2],[ 3]]),         ([4],[[1],[-5],[ 2],[ 3]]),
 ([5],[[1],[ 4],[ 3],[ 2]])]         ([5],[[1],[ 4],[ 2],[ 3]])]
```

```
[([1],[[2],[ 3],[ 4],[ 5]]),        [([1],[[2],[ 3],[ 4],[ 5]]),
 ([2],[[1],[-3],[-4],[-5]]),         ([2],[[1],[-3],[-4],[-5]]),
 ([3],[[1],[ 2],[-5],[-4]]),         ([3],[[1],[ 2],[-4],[-5]]),
 ([4],[[1],[ 2],[-5],[ 3]]),         ([4],[[1],[ 2],[ 3],[-5]]),
 ([5],[[1],[ 2],[ 4],[ 3]])]         ([5],[[1],[ 2],[ 3],[ 4]])]
```

Are there possible reorientations, relabelings, and sign reversals that show the equivalence of all these eight hyperline sequences? First, we provide some Haskell code when we deal with hyperline sequences. Relabeling is a straightforward matter as follows.

Relabeling hyperline sequences in rank 3

```
relabelHypR3::(Int->Int)->[(Tu,OM2)]->[(Tu,OM2)]
relabelHypR3 pi om3 = map (relab pi) om3

relab::(Int->Int)->(Tu,OM2)->(Tu,OM2)
relab pi (om1,om2) = (rel pi om1,map (rel pi) om2)

rel::(Int->Int)->[Int]->[Int]
rel pi om1 = map(\x->(signum x)*(pi(abs x))) om1
```

The problem is that the normalized version of our hyperline sequences in general disappears. The following function `normSeqs` returns the corresponding normalized version in the uniform case.

Find normalized uniform hyperline sequences in rank 3

```
normSeqs::[(Tu,OM2)]->[(Tu,OM2)] ; normSeqs []=[]
normSeqs (l:ls) = sort ([normSeq l]++normSeqs ls)
normSeq::(Tu,OM2)->(Tu,OM2)
normSeq (om1,om2)
  |head om1<0=normSeq([-(head om1)],reverse om2)
  |otherwise =(om1, findNorm om2)
findNorm::[[Int]]->[[Int]]
findNorm om2@(c:cs)
  | head c ==  minimum (map abs (concat om2)) =(c:cs)
  | otherwise = findNorm (cs++[[-(head c)]])
```

The sign reversal for hyperline sequences is done with `signRevR3`.

Sign reversal for uniform hyperline sequences in rank 3

```
signRevR3::[(Tu,OM2)]->[(Tu,OM2)]
signRevR3 om3 = map (signRevRow) om3
signRevRow:: (Tu,OM2)->(Tu,OM2)
signRevRow ([i],om2) = ([negate i],om2)
```

Reorientation for all elements of a given set can be studied with the following function `reorSetR3`.

Reorientation of uniform hyperline sequences in rank 3

```
reorSetR3::[Int]->[(Tu,OM2)]->[(Tu,OM2)]
reorSetR3 [] om3 = om3
reorSetR3 (el:els) om3 = reorSetR3 els (reorElR3 el om3)
reorElR3::Int->[(Tu,OM2)]->[(Tu,OM2)]
reorElR3 el om3 = map (reorRowR3 el) om3
reorRowR3::Int->(Tu,OM2)->(Tu,OM2)
reorRowR3 x (tu,(e:es))
 |x'elem'map abs tu
       =(tu,[e]++map(map negate)(reverse es))
 |x'elem'map abs(concat es)
       =(tu,[e]++map(reorR1 x) es )
 |x'elem'map abs e
       =(tu,[map negate(reorR1 x e)]++map(map negate)es)
reorR1::Int->[Int]->[Int] ; reorR1 el [] = []
reorR1 el (x:xs) |el== abs x = [-x]++xs
                 |otherwise = [ x]++(reorR1 el xs)
```

Now we are ready to compare two uniform hyperline sequences and we can find out how the second can be obtained from the first. The relabeling is given by the first component of the result, the second component tells us which elements have to be reorientated, and the final component tells us whether a sign reversal is necessary. Everything is done by the function `checkOMPair`.

Equivalence check of a pair of oriented matroids in rank 3

```
checkOM3Pair::[(Tu,OM2)]->[(Tu,OM2)]->[([Int],[Int],Bool)]
checkOM3Pair om3A om3B
=  concat[[(s1,s2,sr)]
              |s1 <- permsSet [1..5],
               s2 <- reorLists 5,
               sr <- [True,False],
               sr&& (nS (sR(r (p s1)(rS s2 om3A)))==om3B)]
  ++concat[[(s1,s2,sr)]
              |s1 <- permsSet [1..5],
               s2 <- reorLists 5,
               sr <- [True,False],
               not sr &&(nS (r (p s1)(rS s2 om3A)) ==om3B)]
  where r  = relabelHypR3; p = perm
        rS = reorSetR3
        sR = signRevR3
        nS = normSeqs
```

Finally, we use `[head(checkOM3Pair (rest!!i) (rest!!0))` `|i<-[1..7]]` We compare the last seven hyperline sequences with the first. The result is the following. All these hyperline sequences lie within the same equivalence class with respect to relabelings, reorientations, and sign reversal. Permutations are given by the first component. A sign reversal did not occur.

How to transform the last hyperline sequences into the first

```
[(([1,2,5,4,3],[3,4,5]  ,True),
  ([1,3,2,5,4],[2,3]    ,True),
  ([1,3,4,5,2],[1,2,3,4],True),
  ([1,3,4,5,2],[1,5]    ,True),
  ([1,3,2,5,4],[4,5]    ,True),
  ([1,2,5,4,3],[2]      ,True),
  ([1,2,3,4,5],[1]      ,True)]
```

The result of our investigation is that there is essentially only one uniform oriented matroid with five elements. That oriented matroid has a matrix representation; we can use a regular 5-gon. Its corresponding great circle arrangement can be represented in the projective plane. When we do not label the circles, we have a model for the equivalence class under relabelings. When we do not orientate the circles, we have a model for the equivalence class under reorientations. The change of a sign reversal is just a way of looking at the model. When we finally do not distinguish representations of this model under homeomorphic transformations of the projective plane, we see the cross cap model of the projective plane shown in Figure 3.5 with its five closed curves that pairwise meet precisely once where they cross, as a representative of the only uniform oriented matroid with five elements.

The projective plane cannot be embedded in Euclidean three-space without self-intersections but there are ways to show it, for example, as a Boy surface or as a cross cap model. The latter has the drawback of a singular point (on top) that

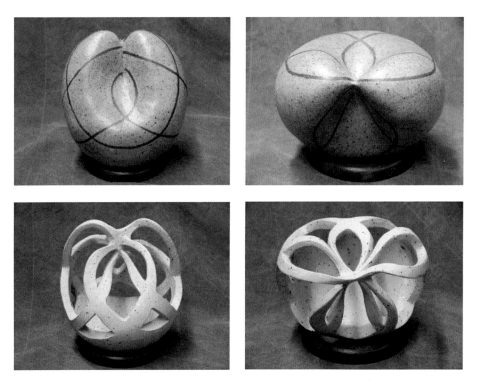

Figure 3.5   Pottery models of the projective plane (cross cap version) with five closed curves intersecting pairwise precisely once, front elevation (left) and top view (right)

has to be identified with a corresponding point below that point, see for example Fomenko, 1994.

This point is not an element for our curves. In order to understand the cross cap model, we use topological deformations step by step. We lift a circular disc, with antipodal boundary points identified, along its "boundary." We use a zip, finally ending up in a vertical line segment, describing the identification of antipodal boundary elements. The endpoints of this line segment have to be identified. In Figure 3.5 we see two pottery models of the cross cap model of the projective plane. The upper model shows five closed curves on the surface. In the lower model, the five curves have been cut off. In both cases, we have a front elevation on the left and a top view on the right. We conclude by repeating: up to topological transformations, the five curves arrangement in the projective plane is unique. This has just been proven by the previous Haskell programs.

3.7 Axiom systems, an overview

If mathematicians are faced with oriented matroids for the first time, they are often irritated if they cannot immediately see a precise definition. We repeat that a particular definition can have certain drawbacks. It shows only one particular aspect of the whole story. There are so many completely different ways to characterize oriented matroids that only a cascade of different definitions, together with experience of how to work with them, would give the right impression of what the theory of oriented matroids is about.

Now we are going to list possible systems of axioms for oriented matroids. In the first four cases, *sphere system axioms*, *hyperline sequences axioms*, *chirotope axioms*, and *circuit axioms*, we will write down the precise definitions later. We do not expect that all readers immediately understand oriented matroids from these definitions via axioms, although the concepts introduced in the previous chapter can be of some help.

We will emphasize in the next section the importance of our first choice of axioms, that is, sphere system axioms, under aspects of invariant theory. Here, the geometric picture gives an essential insight into the theory of oriented matroids.

The second system of axioms, *hyperline sequences axioms*, has the advantage of keeping some of the geometric structure of the sphere system approach. However, the main advantage of this approach lies in its algorithmic applicability, which we will use in later chapters. Its compact data structure is also noteworthy.

The third system of axioms, *chirotope axioms*, seems to be better known than the previous two. An essential proof of equivalence of the first three concepts will follow later in this chapter. We do this in the rank 3 case, where the proof is elementary. We have added a fourth system of axioms, *circuit axioms*, that have played an essential role from the very beginning of oriented matroid theory, see Bland and Las Vergnas, 1978. It shows the surprisingly different flavors in the variety of the following ten cryptomorphic ways to define oriented matroids. We did not aim to list all possible ways. The specialist might miss the tope axioms or Perez Fernandez da Silva's characterization of oriented matroids, see Perez Fernandez da Silva, 1995 and Handa, 1990.

- SPHERE SYSTEMS

  We can define oriented matroids as topological sphere systems, that is, we generalize the oriented great circle arrangements to topological ones and to higher dimensions, see Folkman and Lawrence, 1978.

- HYPERLINE SEQUENCES

  We can define oriented matroids as abstract hyperline sequences, that is, we generalize the concept of hyperlines of point sets to generalized abstract hyperlines, see Bokowski, Mock, and Streinu, 2001, and Bokowski, King, Mock, and Streinu, 2005.

- CHIROTOPES

  We can define oriented matroids as chirotopes, that is, we generalize the concept of a chirotope of a matrix to a combinatorial definition that uses Grassmann–Plücker relations, see Gutierrez Novoa, 1965.

- CIRCUIT AXIOMS

  We can define oriented matroids as a set of abstract Radon partitions via circuit axioms.

- COCIRCUIT AXIOMS

  We can define oriented matroids as a set of abstract hyperplanes via cocircuit axioms.

- ORDER FUNCTIONS

  We can define oriented matroids via order functions, see Kalhoff, 1996, and Jaritz, 1996.

- ALGEBRAIC VARIETIES

  We can use algebraic varieties to define matroids and oriented matroids. After introducing a corresponding set of polynomials, it turns out that matroids and oriented matroids differ precisely by the underlying field, see Bokowski, Guedes de Oliveira, and Richter-Gebert, 1991.

- HULL SYSTEMS

  For characterizing oriented matroids as hull systems, see Knuth, 1992.

- PETRIE POLYGONS

  We can characterize oriented matroids in rank 3 via Petrie polygons. In this way we see oriented matroids as a generalization of Platonic solids, that is, we carry over the cell decomposition properties of the Platonic solids on the sphere to cell decompositions in the projective plane. This possibility has been described in Bokowski, Roudneff, and Strempel, 1997, see also Bokowski and Pisanski, 2005.

- COVECTOR AXIOMS

  We have covector axioms for characterizing oriented matroids.

3.8 Three selected axiom systems in rank 3

In this section, we take three examples out of many possibilities for defining an *oriented matroid*, that is, a generalized equivalence class of matrices. We remain mainly in the rank 3 case. Especially, the sphere system axioms for oriented matroids require in the higher dimensional case topological notations that the reader might want to skip. A profound understanding of hyperline sequences in higher dimensions also requires some familiarity with higher dimensional reasoning. Therefore, we have chosen a far more gentle introduction that starts with the rank 3 case. Our introduction to the abstract general case benefits from what we have prepared in the previous chapter. In a sense we repeat aspects from Chapters 1 and 2 in a condensed mathematical framework in this section. We hope that after all previous considerations, the reader is now ready to work with abstract definitions.

The oldest approach to oriented matroids is that of a sphere system where we have a finite set of oriented closed simple curves on the 2-sphere as a generalization of the oriented great circle concept of the last chapter. We introduce this in the next section.

3.8.1 Sphere systems and Klein's Erlanger program

Among the many equivalent ways to describe an oriented matroid, our first definition of an oriented matroid as a sphere system is in the spirit of Felix Klein's Erlanger Programm. We are going to compare oriented matroids with edge graphs of polytopes in Euclidean space. A polytope is the convex hull of a finite point set in Euclidean space. It can also be seen as a topological ball in Euclidean space.

The *edge graph of the polytope* contains information about the cell decomposition of the space defined by the set of oriented supporting hyperplanes given by the facets of the polytope. The edge graph is invariant under homeomorphic transformations of the underlying space, that is, the graph can be detected even if the edges of the polytope have become curves under homeomorphic transformation. We get interesting example classes of graphs derived from polytopes and point sets.

We generalize the above concept of the edge graph of the polytope. We consider the most general information of the cell decomposition of the space defined by the set of oriented supporting hyperplanes given by the facets of the polytope which is invariant under a homeomorphic transformation of the underlying space, that is, the information of the cell decomposition that can be detected even if the supporting hyperplanes of the polytope have become topological ones under a homeomorphic transformation. This information is called the *oriented matroid of*

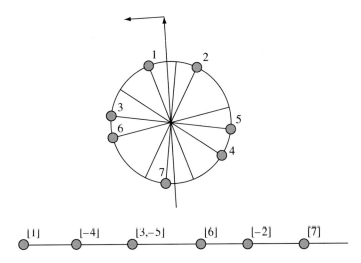

Figure 3.6   Rank 2 example with standard representation

the polytope and we get interesting example classes of oriented matroids derived from polytopes and point sets.

As mentioned before, we look at the rank 3 version first. We start by describing a useful representation of great circle and affine line arrangements. We call it the standard representation for great circle and line arrangements.

Choose an oriented great circle c_i of the oriented great circle arrangement on the sphere S^2, compare Section 1.5, especially Figure 1.7. Orientate the plane spanned by c_i so that its positive side lies to the left if we walk around the circle in the given direction and if we look from the outside. Let A be the oriented plane and A^+ and A^- its two induced open half-spaces. Do an orthogonal projection from the closed hemisphere $S^2 \cap (A^+ \cup A)$ onto A. The resulting planar picture (an oriented circle with oriented arcs inside) will be called the **standard representation** of the oriented great circle arrangement C_n with equator c_i. From the standard representation we can always recover the whole oriented great circle arrangement on the sphere: perform an orthogonal projection in reverse onto the closure of the hemisphere $S^2 \cap A^+$, to obtain oriented half-circles, then by taking antipodal points, complete them to great circles.

We now recommend an exercise to deepen the understanding of the previous concepts.

- Find the standard representation of the example in Figure 3.7.
- Determine the chirotope for this example.
- Find the hyperline sequences for this example.

According to our discussion in Chapters 1 and 2, the standard representation of the oriented great circle arrangement $C_n = \{c_1, \ldots, c_n\}$ with equator c_i can also be

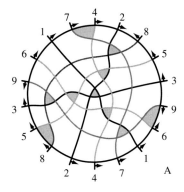

Figure 3.7   A representation of a rank 3 oriented matroid

viewed as a representation of an oriented line arrangement $L_{n-1}^{T} = \{l_1, \ldots, l_n\} \setminus \{l_i\}$ with $n - 1$ elements. If we forget all orientations and extend the tangent plane T, compare Figure 1.11, of our sphere to its projective plane, the standard representation also corresponds to an arrangement of n projective lines, the ith element being the *line at infinity* of T. The standard representation, identifying antipodal points on the circle c_i, defines a cell decomposition of the projective plane induced by the n lines. Note that, in the projective setting, any pair of lines crosses exactly once. We will use the standard representation in two ways: as the projective model, for the cell decomposition properties and incidence properties of its lines and as the sphere system model (as the double covering of the projective plane), for oriented objects.

Definition 3.1 *A* **pseudoline in the projective plane** *is the image of a projective line under a homeomorphic transformation of the projective plane. A* **pseudoline arrangement** $\mathcal{A}$ **in the projective plane** *is a finite ordered set of pseudolines, each pair of which crosses exactly once. We exclude the case if all pseudolines have one point in common.*

The concept of a pseudoline arrangement goes back to Levi, 1926. We have an interesting contribution from Ringel, 1956, and Grünbaum's research monograph from 1972. Let G^T be the group of homeomorphic transformations of the projective plane of T. For an arrangement $\mathcal{A}$ we have the equivalence class of arrangements $cl(\mathcal{A}) := \{\mathcal{A}' | \mathcal{A}' = t\mathcal{A}, t \in G^T\}$. We always consider pseudoline arrangements $\mathcal{A}$ as representatives of their equivalence class $cl(\mathcal{A})$.

We now come back to the sphere S^2 as a double covering of the projective plane. As we use the transition from the standard representation back to the sphere as before, the pseudolines become centrally symmetric, simple, closed curves on the sphere which we call *pseudocircles*. Any pair of pseudocircles crosses in a pair of antipodal points on the sphere (i.e., exactly once in the projective sense). But now we can introduce orientations for all elements to obtain an **arrangement of oriented pseudocircles**. By abuse of terminology, we will refer to our object as an **arrangement of oriented pseudolines** if we want to emphasize the incidence properties inherited from the projective setting and the orientations from the spherical setting. The standard representation has the advantage of showing both these properties.

Definition 3.2 *The* **oriented matroid of rank 3 associated to an arrangement of** *n* **oriented pseudolines** *is its equivalence class with respect to homeomorphic transformations of the projective plane.*

The oriented pseudoline arrangement is called *simple*, *uniform*, or *in general position*, if no more than two pseudolines cross at a point. The example in Figure 3.7 shows a uniform oriented matroid in rank 3 in a standard representation. The marked triangles play a special role. We can change the orientation of such a triangle by moving a pseudoline determined by an edge of such a triangle across the vertex that was not incident with that pseudoline. Clearly, we obtain a new oriented matroid this way. These triangles are, therefore, called mutations. The described transition has also been called a *Reidemeister move* in the literature.

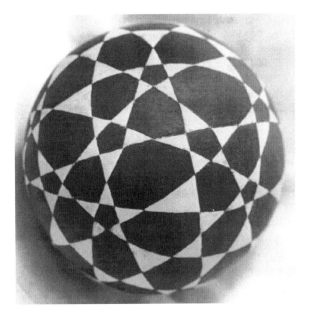

Figure 3.8   Pseudoline arrangement on the sphere, pottery model

3.8.2 Chirotopes, orientations of all simplices

Oriented matroids can be introduced as chirotopes.

Definition 3.3 *A rank 3 oriented matroid with n elements given by a chirotope χ is an alternating and anti-symmetric map $\chi: E^3 \to \{-1, 0, +1\}$ such that $|\chi|$ is a matroid and for pairwise different elements i, j, k, l, m the set of sign products*

$$M := \{\chi(i, j, k) \cdot \chi(i, l, m), \quad -\chi(i, j, l) \cdot \chi(i, k, m), \quad \chi(i, j, m) \cdot \chi(i, k, l)\}$$

contains either zero alone or both values -1 and 1 occur simultaneously, $M = \{0\}$ or $\{-1, +1\} \subset M$. We require in addition that for each element i, there is at least one pair (j, k) with $\chi(i, j, k) \neq 0$.

We have discussed the 3-term Grassmann–Plücker relation in Section 2.2 on Page 45. The essential condition here asserts that the signs of the map should not contradict any single 3-term Grassmann–Plücker relation. This would occur if we allow other values for the signs of χ. For the definition of a matroid see Page 11.

3.8.3 Hyperline sequences, algorithmic advantages

The following Section uses parts of the article Bokowski, Mock, and Streinu, 2001. However, an essential building block of the rank 3 equivalence proof between sphere systems and hyperline sequences that admit an abstract sign of determinant function has been replaced by a Haskell program.

We define hyperline sequences as combinatorial abstractions arising from diverse finite collections of geometric objects, that we have encountered in the previous chapters. In particular, we use vector configurations and arrangements of oriented central planes; arrangements of oriented great circles, and configurations of points on the 2-sphere; as well as arrangements of oriented lines and configurations of signed points. It is an advantage to see these geometrical objects in context. They all lead to hyperline sequences in a natural way and they form paradigms for the later abstract extension to general oriented matroids.

Let us repeat these concepts once more, not attached this time to the only example we used in the last chapters. We consider a non-degenerate **vector configuration** in R^3, that is, a finite ordered set $V_n = \{v_1, \ldots, v_n\} \subset R^3, n \geq 3$, $v_i \neq 0, i = 1, \ldots, n$, such that the one-dimensional subspaces generated by $v_i, i = 1, \ldots, n$, are pairwise different and the corresponding $n \times 3$ matrix M with v_i as its ith row vector has rank 3. The vector configuration will be viewed as a representative of the equivalence class of matrices

$$cl_n(M) := \{M' | M' = DM, D = diag(\lambda_1, \lambda_2, \ldots, \lambda_n), \lambda_i > 0, i = 1, \ldots, n\}.$$

A vector configuration V_n induces an **arrangement of oriented central planes** $H_n = \{h_1, \ldots, h_n\}$, via the concept of polar duality. The unoriented plane of h_i is given as the zero space $\{\underline{x} = (x_1, x_2, x_3) \in R^3 | h_i(\underline{x}) = 0\}$ of a linear homogeneous function $h_i(\underline{x}) = v_{i_1}x_1 + v_{i_2}x_2 + v_{i_3}x_3$, $v_i = (v_{i_1}, v_{i_2}, v_{i_3}) \neq 0$. The positive and negative sides of an oriented central plane are the two induced half-spaces $h_i^+ : \{\underline{x} | h_i(\underline{x}) > 0\}$ and $h_i^- : \{\underline{x} | h_i(\underline{x}) < 0\}$.

An arrangement of oriented central planes H_n induces an **arrangement of oriented great circles** $C_n = \{c_1, \ldots, c_n\}$ on the 2-sphere and vice versa. In $\mathbb{R}^3$, an oriented central plane cuts the unit sphere S^2 along a great circle, which we consider to be parameterized and orientated such that, if looking from the outside, the positive half-space lies to its left as the parameter increases.

A vector $\upsilon \neq 0$, $\upsilon \in R^3$ induces a directed line $l_\upsilon = \{\alpha\upsilon | \alpha \in R\}$ through the origin, which intersects the sphere in two antipodal points s_υ (in the direction of v) and $\bar{s}_\upsilon$ (in the opposite direction). A vector configuration V_n induces a **configuration of points on the sphere**, $S_n = \{s_1, s_2, \ldots, s_n\}$, where $s_i = s_{\upsilon_i}$, $i = 1, \ldots, n$. Each point p on the sphere has an associated antipodal point $\bar{p}$.

We carry over the previous polar dual pairs to the affine plane T, viewed as a plane tangent to the 2-sphere. We assume that v_i, $i \in \{1, \ldots, n\}$ is neither parallel nor orthogonal to the plane T.

The great circle parallel to T defines two open hemispheres. One of them, called the *upper hemisphere*, contains the tangent point of T. An oriented great circle c_i induces an oriented half-circle in this upper hemisphere. This projects onto an oriented straight line $l^T(c_i)$ in the plane T via radial projection and, vice versa, any oriented straight line in T defines an oriented great circle on S^2. An arrangement of oriented great circles C_n induces an **arrangement of oriented lines** in the affine plane $L_n^T = \{l_1, \ldots, l_n\}$, where $l_i := l^T(c_i)$.

The same transition from the sphere S^2 to the plane T leads from a point configuration on the sphere to a signed point configuration in the affine plane. We define $sp^T(s_i)$ to be a pair of a signed index and a point $p_i \in T$ obtained via radial projection from s_i, as follows. A point s_i on the upper hemisphere maps to the pair $sp^T(s_i) = (i, p_i)$, $i \in \{1, \ldots, n\}$ and a point s_i on the lower hemisphere maps to a pair $(\bar{i}, p_{\bar{i}})$ and $p_{\bar{i}} := p_i \in T$. For an element i of our index set E, we find for the negative element, $-i$, often the notation $\bar{i}$ in the literature that we use here as well. In general we prefer the negative sign, especially, if we have a Haskell program in mind. We obtain from $S_n = \{s_1, \ldots, s_n\}$ a **signed point configuration** $P_n^T = \{sp_1, \ldots, sp_n\}$, with $sp_i := sp^T(s_i)$ and vice versa.

We use $E_n = \{1, \ldots, n\}$, endowed with the natural order, to denote the index set of geometric objects like vectors, planes, great circles, and points on the sphere; lines and points in the Euclidian plane; or of a finite ordered set of abstract elements. The associated *signed index set* $\overline{E}_n = \{1, \ldots, n, \bar{1}, \ldots, \bar{n}\}$ makes it possible to denote orientations or signs of these elements. The $s \mapsto \bar{s}$ operator is required to be an involution: $\bar{\bar{s}} = s$, $\forall s \in \overline{E}_n$.

All of the above ordered sets $V_n, H_n, C_n, S_n, L_n^T, P_n^T$ can be viewed as geometric interpretations of the same equivalence class of matrices $cl_n(M)$. We can reorientate the elements. The reorientation classes are the equivalence classes with respect to reorientation. These reorientation classes are obtained if the numbers $\lambda_i \neq 0$ can also be negative. The *reorientation of a vector* v_i is the vector $v_{\bar{i}} = -v_i$ and the *reorientation of an oriented central plane* is the change in the sign of its normal vector. The *reorientation of an oriented great circle* or of an *oriented line* means replacing it by the same object with reversed orientation. The *reorientation of a signed point* $(i, p_i), i \in \overline{E}_n$ is the signed point $(\bar{i}, p_{\bar{i}})$, $p_i = p_{\bar{i}}$. The *reorientation of an index* i is its replacement by $\bar{i}$. The *relabeling of an ordered set* is given by a permutation of its elements.

We now extract combinatorial information from all the geometric sets defined above. We will work *only* with signed subsets $q \subset \overline{E}_n$ which do not contain simultaneously both an element i and its negation $\bar{i}$. If $q \subset \overline{E}_n$, we define $\bar{q} = \{\bar{s} | s \in q\}$. The unsigned support $supp(q) \subset E_n$ of $q \subset \overline{E}_n$ is obtained by ignoring all the signs in q. A *signed partition* of E_n is a signed set $I = I^+ \cup I^-$ with $I^+, \overline{I^-} \subset E_n$, $I^+ \cup \overline{I^-} = E_n$. The following abstract definition will be useful when we later extend the geometrical examples of this section to a purely combinatorial level.

Definition 3.4 *A **hyperline sequence** hs_i over $\overline{E}_n, i \in \overline{E}_n$, with half-period length k_i is a pair $hs_i = (i, \pi_i)$, where π_i is a double infinite sequence $\pi_i = (q_j^i)_{j \in \mathbb{Z}}$ with $q_j^i \subset \overline{E}_n \setminus \{i, \bar{i}\}$, $q_j^i = \overline{q_{j+k_i}^i}$, $\forall j \in \mathbb{Z}$, $supp(\bigcup_{j \in \mathbb{Z}} q_j^i) = E_n \setminus supp(\{i\})$, where the unsigned supports of $q_1^i, \ldots, q_{k_i}^i$ are mutually disjoint. We consider $hs_i = (i, \pi_i)$ and $hs_{\bar{i}} = (\bar{i}, \pi_{\bar{i}})$ to be equivalent if π_i is obtained from $\pi_{\bar{i}}$ by reversing the order.*

```
[([1], [[2],[ 3],[ 4],[ 5],[ 6],[ 7]]),
 ([2], [[1],[-3],[-4],[-5],[-6],[-7]]),
 ([3], [[1],[ 2],[-4],[-5],[-6],[-7]]),
 ([4], [[1],[ 2],[ 3],[-5],[-6],[-7]]),
 ([5], [[1],[ 2],[ 3],[ 4],[-6],[-7]]),
 ([6], [[1],[ 2],[ 3],[ 4],[ 5],[-7]]),
 ([7], [[1],[ 2],[ 3],[ 4],[ 5],[ 6]])]
```

The reader should confirm that the hyperline sequences definition captures the structure we encountered before, for example, in our planar example in Section 3.3. The infinite sequence concept was chosen to make the definition independent of the normalization that the smallest element comes first with positive sign. We have written above these hyperline sequences in Haskell code.

In the particular case if all the q_j^is are one-element subsets, the sequence is said to be in *general position, simple,* or *uniform,* and we replace the sets q_j^i with their elements. In this case, any half period of π_i is a signed permutation of $E_n \setminus supp(\{i\})$. In general, we have an additional ordered partition into pairwise disjoint subsets of the signed elements. An infinite sequence π_i in a hyperline sequence $hs_i = (i, \pi_i)$ can be represented by any half period, that is, by any k_i consecutive signed sets $q_{t+1}^i, \ldots, q_{t+k_i}^i, q_{t+j}^i \subset \overline{E}_n \setminus \{i, \overline{i}\}, t \in Z$.

We obtain the *normalized representation* $hs_r = (r, \pi_r)$ of a hyperline sequence $hs_i = (i, \pi_i)$ by first choosing $(r, \pi_r) := (i, \pi_i)$ if $i \in E_n$ or $(r, \pi_r) := (\overline{i}, reverse(\pi_i))$ if $\overline{i} \in E_n$, and then choosing the half-period of π_r starting with the set $q_j^r \subset \overline{E}_n$ containing the smallest positive element. To a **signed point config-** **uration** $P_n^T = \{(i, p_i) | i \in I\}$ (obtained from a vector configuration as described above) we associate a set $HS(P_n^T) = \{hs_1, \ldots, hs_n\}$ of n hyperline sequences $hs_i = (i, \pi_i)$ over $\overline{E}_n$. The sequence π_i, denoted by a half-period $q_1^i, q_2^i, \ldots, q_{k_i}^i$, with $q_j^i \subset \overline{E}_n \setminus \{i, \overline{i}\}$, corresponds to the signed point $(i, p_i) \in P_n^T$. It is obtained by rotating an oriented line in counterclockwise (ccw), or in clockwise (cw), order around p_i if $i \in E_n$, or if $\overline{i} \in E_n$, respectively, and looking at the successive positions where it coincides with lines defined by pairs of points (p_i, p_j) with $p_j \neq p_i$. If P_n^T is not in general position, several points may become simultaneously collinear with the rotating line, and they are recorded as a set q_k^i. If the point p_j of the signed point (j, p_j) is encountered by the rotating line in positive direction from p_i, it will be recorded as the index j, otherwise as the negated index $\overline{j}$. The whole sequence is recorded in the order induced by the rotating line, and an arbitrary half-period is chosen to represent it.

Definition 3.5 *The rank 3* **oriented matroid induced by hyperline sequences** **associated to a signed point configuration** $P_n^T = \{(i, p_i) | i \in I\}$, *where I is a signed partition of E_n, is $HS(P_n^T) = \{hs_i = (i, \pi_i) | i \in I\}$ as described above. We identify $HS(P_n^T)$ with $\{(\overline{i}, \pi_i) | i \in I\}$.*

Later on we will often consider sets of hyperline sequences that *admit an abstract sign of determinant function.* When this is clear from the context, we will use hyperline sequences in this restricted sense.

Note that if the orientation of the plane T is reversed, all the sequences are reversed. The identification in the previous definition makes the notion of hyperline sequences independent of the chosen orientation of the plane T.

Remark. If we start with a set of vectors V_n and two admissible tangent planes T and T', by radial projection, we obtain two sets of signed planar points P_n^T and $P_n^{T'}$. The reader can verify that our definition ensures that the resulting hyperline sequences $HS(P_n^T)$ and $HS(P_n^{T'})$ will coincide. This allows for a definition of hyperline sequences associated to any of the previously considered geometric ordered sets: vectors, oriented central planes, etc.

We explicitly describe this in two cases. Consider an arrangement $C = \{c_1, \ldots, c_n\}$ of n oriented great circles on the sphere S^2. To each circle c_i associate a hyperline sequence by recording the points of intersection (ordered according to the orientation of the circle c_i) with the remaining oriented circles. An index j is recorded as positive (respectively negative) if the circle c_j crosses c_i from left to right (respectively right to left). When we use an arrangement of oriented lines $L_n^T = \{l_1, \ldots, l_n\}$, it induces a set of n hyperline sequences $HS(L_n^T)$ as follows: for each line l_i record the points of intersection with the other lines (ordered according to the orientation of the line). Each element j is signed: positive if line l_j crosses l_i from left to right, negative otherwise.

Hyperline sequences of configurations and arrangements of the last section store the signs of determinants of 3×3 submatrices of the matrix M of a corresponding vector configuration $V_n = \{v_1, \ldots, v_n\} \subset R^3, n \geq 3$ $v_i \neq 0$, $i = 1, \ldots, n$. This is an invariant for all matrices $M' \in cl_n(M)$. Let i, j, k be three distinct signed indices in $\overline{E}_n$. Let [i,j,k] be the determinant of the submatrix of M with row vectors v_i, v_j, v_k. If j and k appear within the same set q_k^i of π_i, we have $\text{sign}[i, j, k] = 0$. If j and k occur in this order in some half-period of π_i, we have $\text{sign}[i, j, k] = +1$ and $\text{sign}[i, j, k] = -1$ otherwise. The sign of the determinant $\chi(ijk) := \text{sign}[i, j, k]$ is independent of the chosen half periods and compatible by alternation $\chi(ijk) = \chi(jki) = \chi(kij) = -\chi(ikj) = -\chi(kji) = -\chi(jik)$ and anti-symmetry $\chi(\bar{i}jk) = -\chi(ijk)$.

Given an abstract set of hyperline sequences, we choose its corresponding normalized form and define $\chi : \overline{E}_n^3 \to \{-1, 0, +1\}$, (partially) by: $\chi(ijk) := 0$, if j and k appear within the same set q_s of π_i, for i in E_n, j, k in $\overline{E}_n$, $j \neq k$, $\chi(ijk) := +1$, if j and k occur in this order in π_i, and $\chi(ijk) := -1$, if j and k occur in the reversed order in π_i.

Extending this partial definition of χ by alternation and anti-symmetry, the value of $\chi(ijk)$ for $0 < i < j < k$ is obtained either directly, by the above rule applied to each of the three hyperline sequences, or via alternation and anti-symmetry. When these three values for $\chi(ijk)$ are compatible in all cases, we say that *the set of hyperline sequences admits an abstract sign of determinant function.*

Definition 3.6 *A rank 3 oriented matroid with n elements given by hyperline sequences is a set of hyperline sequences $\{(i, \pi_i)|i \in I\}$ over $\overline{E}_n$ which admit an abstract sign of determinant function. The oriented matroid is uniform when all hyperline sequences are uniform.*

Let us summarize that we have defined oriented matroids of rank 3 in essentially three different ways. Each case was a generalization of vector configurations and of arrangements of oriented central planes; of arrangements of oriented great circles and of configurations of points on the 2-sphere; and of arrangements of oriented lines and of configurations of signed points.

We have defined rank 3 oriented matroids with n elements associated to an arrangement of n oriented pseudolines in Definition 3.2. We have defined a rank 3 oriented matroid with n elements given by a chirotope in Definition 3.3. Finally, we have defined rank 3 oriented matroids with n elements given by hyperline sequences in Definition 3.4. We now have to prove that these three concepts coincide.

3.9   Axiom equivalence in rank 3

3.9.1   *Chirotopes and hyperline sequences*

In this section, we show that hyperline sequences admitting an abstract sign of determinant function of Definition 3.6 and chirotopes, compare Definition 3.3, define the same class of objects. We have worked out how to convert chirotopes to hyperline sequences and vice versa in the uniform case with our functions `chi2HypR3` and `hyp2ChiR3`. The corresponding ideas lie behind the following equivalence proofs. We establish the equivalence between these two systems of axioms, according to Bokowski, Mock, and Streinu, 2001.

Theorem 3.7   *There is a one-to-one correspondence between rank 3 oriented matroids given by hyperline sequences and those defined by chirotopes.*

Proof. Let the rank 3 oriented matroid with n elements be given by hyperline sequences. We will show that the abstract sign of determinant function fulfills the chirotope condition: $\forall i, j, k, l, m$, pairwise different,

$$M := \{\chi(i,j,k) \cdot \chi(i,l,m), -\chi(i,j,l) \cdot \chi(i,k,m), \chi(i,j,m) \cdot \chi(i,k,l)\} = \{0\}$$
$$\text{or } \{-1, +1\} \subset M,$$

which is invariant under permutation of the elements j, k, l, m and reorientation of all its five elements i, j, k, l, m. In the ith hyperline sequence, we can assume that the elements j, k, l, m occur in that order. Elements belonging pairwise to different q_s, imply $\chi(i,j,k) = \chi(i,l,m) = \chi(i,j,l) = \chi(i,k,m) = \chi(i,j,m) = \chi(i,k,l) = 1$. If all the elements belong to the same q_s, we have $\chi(i,j,k) = \chi(i,l,m) = \chi(i,j,l) = \chi(i,k,m) = \chi(i,j,m) = \chi(i,k,l) = 0$. We use $q(t)$ for the set containing element t and $q(s) < q(t)$ says that $q(s)$ comes before $q(t)$ in the chosen half-period of i. The remaining cases are now $q(j) < q(k) < q(l) = q(m)$, $q(j) < q(k) = q(l) < q(m)$, $q(j) = q(k) < q(l) < q(m)$, $q(j) = q(k) = q(l) = q(m)$,

$q(j) = q(k) = q(l) < q(m)$, and $q(j) = q(k) < q(l) = q(m)$. We easily see that the chirotope condition holds in all these cases.

In order to prove the reverse direction, let the chirotope be given. We have to show that we can construct an oriented matroid induced by a set of hyperline sequences. The abstract sign of determinant function will be the function χ of the chirotope. During the construction process, we confirm that all images of χ are compatible with the hyperline sequence structure. For the given element i we have at least one pair $(a, b) \in \overline{E}_n^2$ with $\chi(i, a, b) = 1$. We start to construct a half-period of the ith hyperline by sorting these two elements in the correct order beginning with $a \in \overline{E}_n$ followed by $b \in \overline{E}_n$. Now we use induction. Using for the next element $k \notin q(a)$ k or $\overline{k}$ depends on $\chi(i, k, a)$. Using for the next element $k \in q(a)$ k or $\overline{k}$ depends on $\chi(i, k, b)$. The first element k not belonging to $q(a)$ and $q(b)$ gets its unique position by $\chi(i, k, b)$. For additional insertions we have to show compatibility. Assume that $k - 1$ elements have already been sorted in the correct way, forming the ordered sets $q_1 < q_2 < \cdots < q_t, t \geq 3$. We consider all signs $\chi(i, k, x)$. We insert the kth element which was not used thus far. We observe first that $\chi(i, k, x)$ is the same for all $x \in q_s$ for some s. We show this for $q_s \neq q(a)$. (Otherwise $q(b)$ can be used instead.) For $x_1, x_2 \in q_s$ we have $\chi(i, x_1, x_2) = 0$ and $\chi(i, a, x_1) = \chi(i, a, x_2) = 1$. The chirotope property gives us either $\chi(i, k, x_1) = \chi(i, k, x_2) = 0$, that is, $k \in q_s$ or $\chi(i, k, x_1) = \chi(i, k, x_2) \neq 0$. The sorting of the elements $x \in \overline{E}_n$ in the ith half-period is a sorting of its classes $q(x)$. Assuming $q(a) < q(x) < q(y)$ and $q(a) < q(y) < q(z)$ $a, x, y, z \in \overline{E}_n$, we have $\chi(i, a, x) = 1$, $\chi(i, y, z) = 1$, $\chi(i, a, y) = 1$, $\chi(i, a, z) = 1$, $\chi(i, x, y) = 1$. The chirotope property implies $q(x) < q(z)$, that is, the ordering is always compatible: for $k = x$ we find that there is either a least upper bound or no upper bound, for $k = z$ we find that there is a largest lower bound and thus insert k in the sequence. For all i we can construct the corresponding half-period in accordance with the chirotope function which serves as the abstract sign of determinant function.

3.9.2 Sphere systems and hyperline sequences

Understanding the inductive structure of the proof of the following theorem helps to understand later the algorithmic generation of extensions of oriented matroids.

Theorem 3.8 *[Topological representation theorem, rank 3] There is a one-to-one correspondence between rank 3 oriented matroids, given by hyperline sequences, and equivalence classes of oriented pseudoline arrangements in the projective plane.*

Proof. We use mainly the proof from Bokowski, Mock, and Streinu, 2001. We assume that the easy direction, starting with an oriented pseudoline arrangement and finding its corresponding hyperline sequences, causes no difficulties for the reader. So, we start with a rank 3 oriented matroid with n elements given by hyperline sequences in normalized form. We are going to construct an arrangement of oriented pseudolines with line 1 being the line at infinity. We prove the uniform case by induction, showing that if an arrangement of $n - 1$ oriented pseudolines has been constructed, it is possible to insert the nth oriented pseudoline in a manner compatible with the given hyperline sequences.

Part A. The case $n \leq 5$. Here the theorem is true because we have precisely one example for $n = 5$. We have shown this within this chapter. For the proof of the topological representation theorem in rank 3 the case $n = 5$ forms a fundamental base step for the induction.

Part B. We apply induction to get an arrangement PL_{n-1} of $n - 1$ oriented pseudolines with pseudoline 1 as the line at infinity, whose set of normalized hyperline sequences is obtained by removing the element n from each sequence and deleting the nth sequence. Using the position of element n in each of the original hyperline sequences, we mark $n - 1$ points, labeled with unordered pairs of indices (i, n), $i = 1, \ldots, n - 1$ (denoted for simplicity as ni) on the existing pseudolines $1, \ldots, n - 1$.

The nth hyperline sequence defines an ordering of these points $n : 1$, $k_1, k_2, \ldots, k_{n-2}$. Here and in what follows we understand the index j modulo $(n - 1)$ and $k_0 = 1$. To prove the inductive step, it suffices to show that the following three conditions hold.

(i)   We can join any two consecutive points nk_j and $nk_{j+1}, j = 0, \ldots, n - 2$ with a pseudoline segment such that the open segment does not meet any of the already existing pseudolines.
(ii) The resulting curve $\mathcal{K}$ obtained from all these segments together with the already existing pseudolines form a pseudoline arrangement PL_n, that is, $\mathcal{K}$ is a simple closed curve which crosses each pseudoline $j, j \leq n - 1$ at nj and nowhere else.
(iii) The nth pseudoline has a unique orientation.

Proof of (i). We show that any two consecutive points $nk_j, nk_{j+1}, j = 0, \ldots, n - 2$ belong to the same cell of the arrangement PL_{n-1}, that is, they are not separated by any of the existing pseudolines $i, i \neq 1, k_j, k_{j+1}$.

Consider two consecutive points nk_j, nk_{j+1} for an index $j = 0, \ldots, n - 2$ and a pseudoline $i \neq 1, k_j, k_{j+1}$. Applying the induction hypothesis to the restriction of HS to the set of five, respectively four, elements $\{1, k_j, k_{j+1}, i, n\}$ (extensions up to four and five elements are even unique) implies a unique corresponding oriented pseudoline arrangement PL_5, respectively PL_4, up to a homeomorphic transformation of the projective plane. We have four elements, for example, in the special case $j = 0$, because $k_0 = 1$. The ith pseudoline does not separate the points nk_j, nk_{j+1}. Therefore any two consecutive points nk_j, nk_{j+1}, $j = 0, \ldots, n-2$ of the arrangement PL_{n-1}, are not separated by any of the existing pseudolines i, $i \neq 1, k_j, k_{j+1}$. This implies that we can connect two consecutive points nk_j, nk_{j+1} by a pseudoline segment without crossing any of the existing pseudolines.

Proof of (ii). We consider two consecutive points nk_j, nk_{j+1}, $j = 0, \ldots, n-2$ and we pick a point np_j on the open pseudoline segment nk_j, nk_{j+1}, as constructed above. We show that all points nk_i, $i \in \{j+2, \ldots, n-2\}$ are separated from point np_j by the pseudoline k_{j+1} and in a similar way that all points nk_i, $i \in \{1, \ldots, j-1\}$ are separated from point np_j by pseudoline k_j. The argument in both cases is the same, so we prove only the first case (see Figure 3.9). We restrict the hyperline sequences to the set $\{1, k_j, k_{j+1}, i, n\}$ (the cases $1 = k_0$ and $k_{n-1} = 1$ are included), in which we find a unique corresponding pseudoline arrangement, which uses the open pseudoline segment from nk_j to nk_{j+1}. The separation property holds and it carries over to the arrangement with $n-1$ elements. This implies that the closed curve $\mathcal{K}$ above consisting of all pseudoline segments has no self-intersections, and it crosses all other $n-1$ pseudolines just once.

Proof of (iii). On each oriented pseudoline $i \in \{1, \ldots, n-1\}$, we put an arrow A_{ni} at point ni pointing to the right side (or to the left side) of pseudoline i if the sign of element n in the ith hyperline sequence is positive (or negative, respectively). We show that for any two consecutive points nk_j, nk_{j+1}, $j = 0, \ldots, n-2$, the corresponding arrows A_{nk_j}, $A_{nk_{j+1}}$ are compatible, that is, the induced orientation of pseudoline n by A_{nk_j} coincides with that of $A_{nk_{j+1}}$. We restrict the hyperline sequences to the set $\{1, k_j, k_{j+1}, n\}$. Applying the induction hypothesis for each j shows that we have in each case a unique corresponding oriented pseudoline

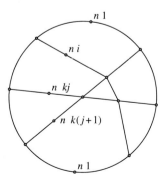

Figure 3.9   Restriction on five elements

arrangement with the desired property. This implies a unique orientation of the *n*th pseudoline in the globally constructed pseudoline arrangement. This concludes the proof by induction that a new pseudoline can be inserted in the uniform case.

The non-uniform case is also proven inductively, by eliminating one degeneracy at a time until we obtain the uniform case. The corresponding proof in the arbitrary rank case avoids this additional induction.

3.10   Non-realizable rank 3 oriented matroids

Levi had already in his 1926 paper mentioned the projective incidence theorem of Pappus as a source for finding non-realizable rank 3 oriented matroids.

This observation holds on a much larger scale. We can use any projective incidence theorem which asserts that given a set of points and a set of lines in the (projective) plane, defined as joins of subsets of these points, three of these lines necessarily meet in one point. When we use orientations for these lines and when we distort one of the three lines into a pseudoline that does not meet the point of intersection, we have defined a non-realizable rank 3 oriented matroid because of that projective incidence theorem.

The finite symmetrical point set on a 2-sphere on the left of Figure 3.10 is known as the Pappus configuration. A central projection from the midpoint of the 2-sphere onto a Euclidean plane touching the 2-sphere in a visible fixpoint of the rotational symmetry of order 3 leads to a corresponding finite point set representation in the Euclidean plane. However, the sphere arrangement is a better model for the discussion of properties of this configuration.

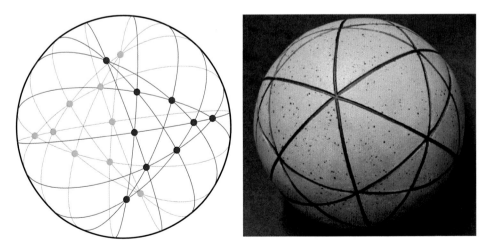

Figure 3.10   Pappus configuration, computer graphic and pottery model

The Pappus configuration leads to the following hyperline sequences.

```
[( [1], [[2],    [4,5], [6,9], [7,8], [3]    ]),
 ( [2], [[3],    [6,7], [8,5], [9,4], [1]    ]),
 ( [3], [[1],    [8,9], [4,7], [5,6], [2]    ]),
 ( [4], [[2,-9], [5,-1], [6],   [3,7], [8]    ]),
 ( [5], [[3,6], [7],   [8,-2], [9],   [1,4] ]),
 ( [6], [[3,-5], [7,-2], [8],   [1,9], [4]    ]),
 ( [7], [[1,8], [9],   [4,-3], [5],   [2,6] ]),
 ( [8], [[1,-7], [9,-3], [4],   [2,5], [6]    ]),
 ( [9], [[2,4], [5],   [6,-1], [7],   [3,8] ])]
```

How does the data structure for this oriented matroid look in other models? We have determined the chirotope with `map snd hyp2ChiR3` from the given hyperline sequences.

Corresponding chirotope of the Pappus configuration

```
[+,+,+,+,+,+,+,-,-,-,-,-,-,0,+,+,+,+,+,+,+,
 +,+,+,0,0,-,-,+,+,+,+,+,+,-,-,-,-,0,-,-,0,
 +,0,+,+,+,+,+,+,+,0,-,-,0,-,-,-,-,-,-,-,-,
 0,+,+,+,+,+,+,+,+,+,+,+,+,+,+,+,+,+,+,+,+]
```

This information is again a reasonably compact data structure, however, geometric ideas seem to be completely hidden.

We obtain all circuits of this oriented matroid when we use the function `chi2C` from Page 57 and when we apply it to the chirotope. The data structure would cover a whole page This is why we refrain from using the circuit or cocircuit information of the oriented matroid for applied purposes.

Circuits of the Pappus example

```
[[+,0,0,-,+,0,0,0,0],[-,0,0,+,-,0,0,0,0],
 [-,0,0,0,0,-,0,0,+],...the data structure
 of all circuits would cover a whole page
 [0,0,0,0,0,+,-,+,-],[0,0,0,0,0,-,+,-,+]]
```

We show in Figure 3.11 a slightly distorted (reorientation class of a) pseudoline arrangement that is no longer representable as a 1-sphere arrangement on the 2-sphere. The proof technique for that from the theory of oriented matroids can be carried over to prove Pappus theorem. Moreover, an investigation of

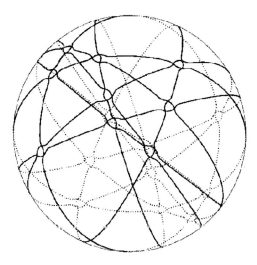

Figure 3.11   Uniform Pappus configuration in the projective plane

non-realizable rank 3 oriented matroids with up to ten elements has led to a systematic overview about projective incidence theorems with up to ten elements, in an unpublished article of Bokowski, Laffaille, and Richter-Gebert, 1989. We repeat, a projective incidence theorem can be used to construct a non-realizable rank 3 oriented matroid. When we have a sieve method that leads to the non-realizable oriented matroids, we have a means to search for projective incidence theorems systematically.

We come back to this topic in Section 10.5 in the context of several other projective incidence theorems.

4

Oriented matroids of arbitrary rank

We view the first three chapters as a preparation for accepting general oriented matroid axioms, and here they come. We first introduce the sphere system model and the chirotope model for oriented matroids of arbitrary rank. The sphere system model is a generalization of Levi's pseudoline concept of 1926 and it allows us to think in geometric terms. Combinatorial properties of geometric objects like convex polytopes, general polyhedra, point configurations, and hyperplane arrangements appear within a general framework in these sphere systems. The surprisingly one-to-one correspondence between the sphere system model of oriented matroids and the chirotope model for oriented matroids allows us to work with signs. An inductive enumeration of combinatorial types of convex polytopes is not available in higher dimensions; however, we can enumerate all chirotopes, thus gaining a superset of all combinatorial types of convex polytopes.

We do not aim to show proofs of equivalence between all of these axiom systems. The proofs are very often not at all easy. For such proofs we refer either to original papers or to the research monograph Björner *et al.*, 1993. The topological representation theorem of Folkman and Lawrence is considered to be one of the foundations of oriented matroid theory; compare also a recent topological representation theorem for matroids, see Swartz, 2002. The new proof of the topological representation theorem for oriented matroids is a simpler version compared with the original one of Folkman and Lawrence that we can find as a highlight in Chapter 5 of the research monograph of Björner *et al.*, 1993. The new proof shows the equivalence of oriented matroids in terms of chirotopes, sphere systems, and hyperline sequences, see Bokowski, King, Mock, and Streinu, 2005. We have seen its basic idea in Chapter 3 where we have presented the rank 3 version. We have simplified one part of the original proof by replacing it with some Haskell code. The higher dimensional case works in a similar way, however, additional concepts from topology have to be introduced. The inductive structure of the proof is the key idea for extension algorithms for generating

111

all oriented matroids of given number of elements and given rank. Such an extension algorithm played a fundamental role in many results in which the algorithm was not even mentioned because it was enough to present the final coordinates. Because of this decisive role, we refer to some C++ code available from `http://juergen.bokowski.de` in the uniform rank 4 case. It was a key step for the result in Bokowski and Guedes de Oliveira, 2000. Later we provide a Haskell program for such an extension algorithm in the rank 5 case for polytopes. This case has many applications and it is much simpler than the most general case.

Now we start to introduce various equivalent axiom systems of oriented matroids. Sphere systems or arrangements of oriented pseudospheres very often provide us with geometric insight of our problems. Chirotopes and especially hyperline sequences have the advantage that the data structure is compact for extension algorithms. Circuits axioms or cocircuit axioms also lead to classical data structures for oriented matroids.

With the chirotope model we investigate a non-trivial example of a 4-dimensional polytope that has no diagonals.

4.1 Sphere systems, oriented pseudospheres

We use the $(r-1)$-dimensional unit sphere $\mathbb{S}^{r-1} \subset \mathbb{R}^r$ and a finite multiset of non-zero vectors $V \subset R^r$. Any vector $v \in V$ yields a pair $\pm v/\|v\|$ of points in $\mathbb{S}^{r-1}$. This is dual to a hypersphere $S_v \in \mathbb{S}^{r-1}$, namely the intersection of $\mathbb{S}^{r-1}$ with the hyperplane $H_v \subset \mathbb{R}^r$ that is perpendicular to v. We choose a fixed orientation on H_v, so that a positive base of H_v together with v is a positive base of $\mathbb{R}^r$. This induces an orientation on S_v. The set of vectors V is dual to an arrangement of oriented hyperspheres in $\mathbb{S}^{r-1}$ that we are going to generalize. The idea is to consider arrangements of oriented embedded topological spheres of codimension 1 in $\mathbb{S}^{r-1}$, intersecting each other like the above hyperspheres.

We formalize this idea. Let $\mathbb{S}^d$ denote the d-dimensional oriented sphere

$$\mathbb{S}^d = \left\{ (x_1, \ldots, x_{d+1}) \in \mathbb{R}^{d+1} \,|\, x_1^2 + \cdots + x_{d+1}^2 = 1 \right\},$$

and let

$$B^d = \left\{ (x_1, \ldots, x_d) \in \mathbb{R}^d \,|\, x_1^2 + \cdots + x_d^2 \le 1 \right\}$$

denote the closed d-dimensional ball. A submanifold N of codimension m in a d-dimensional manifold M is *tame* if any $x \in N$ has an open neighborhood $U(x) \subset M$ such that there is a homeomorphism $\overline{U(x)} \to B^d$ sending $U(x) \cap N$ to $B^{d-m} \subset B^d$.

An *oriented pseudosphere* $S \subset \mathbb{S}^d$ is a tame embedded $(d-1)$-dimensional sphere with a choice of an orientation. Obviously any oriented hypersphere is an oriented pseudosphere. Let $\psi\colon \mathbb{S}^{d-1} \to \mathbb{S}^d$ be an embedding with image S, inducing the desired orientation of S. According to M. Brown, 1962, the image of ψ is tame if and only if ψ can be extended to an orientation preserving embedding

$$\tilde{\psi}\colon \mathbb{S}^{d-1} \times [-1, 1] \to \mathbb{S}^d \text{ with } \psi(\cdot) = \tilde{\psi}(\cdot, 0).$$

The image of an oriented pseudosphere S under a homeomorphism $\phi\colon \mathbb{S}^d \to \mathbb{S}^d$ is again an oriented pseudosphere, since the defining embedding $\phi \circ \psi\colon \mathbb{S}^{d-1} \to \mathbb{S}^d$ can be extended to $\phi \circ \tilde{\psi}\colon \mathbb{S}^{d-1} \times [-1, 1] \to \mathbb{S}^d$. By the generalized Schönflies theorem, $\mathbb{S}^d \setminus S$ is a disjoint union of two open balls whose closures are closed balls. We call the connected component of $\mathbb{S}^d \setminus S$ containing $\tilde{\psi}(\mathbb{S}^{d-1} \times \{1\})$ (respectively $\tilde{\psi}(\mathbb{S}^{d-1} \times \{-1\})$) the *positive side* S^+ (respectively *negative side* S^-) of S.

Let $E_n = \{1, \ldots, n\}$.

Definition 4.1 *[Arrangement of oriented pseudospheres] Let* $S_1, \ldots, S_n \subset \mathbb{S}^d$ *be not necessarily distinct oriented pseudospheres, ordered according to their indices. The ordered multiset* $\{S_1, \ldots, S_n\}$ *is an arrangement of oriented pseudospheres over* E_n *if for all* $R \subset E_n$ *and for all* $i \in E_n$

- $S_R := \mathbb{S}^d \cap \bigcap_{j \in R} S_j$ *is empty or homeomorphic to a sphere,*
- $S_R \not\subset S_i$ *implies* $S_R \cap S_i$ *is a pseudosphere in* S_R, $S_R \cap S_i^+ \neq \emptyset$, $S_R \cap S_i^- \neq \emptyset$.

We use arrangements of full rank, that is, the intersection of all its members is empty.

Definition 4.2 *[Equivalence class=oriented matroid] Two ordered multisets* $\{S_1, \ldots, S_n\}$ *and* $\{S_1', \ldots, S_n'\}$ *of oriented pseudospheres in* $\mathbb{S}^d$ *are equivalent if there is an orientation preserving homeomorphism* $\mathbb{S}^d \to \mathbb{S}^d$ *sending* S_i^+ *to* $(S_i')^+$ *and* S_i^- *to* $(S_i')^-$, *simultaneously for all* $i \in E_n$.

We call the equivalence class of an arrangement of oriented pseudospheres an *oriented matroid*. When we wish to stress the underlying concept, we can also use the notion *sphere system*. Recall what we have said about Klein's Erlanger Programm on Page 95. A sphere system is by definition an invariant under homeomorphic transformations. It is convincing, at least for mathematicians, that we have reached with the concept of a sphere system a satisfying definition of an oriented matroid and not just an artificial concept. When we define for a great sphere arrangement, a cell of maximal dimension to be a spherical polytope, we now have a generalized notion for such a topological cell within a sphere system. The notion of a facet, a cell of lower dimension, and the edge graph of such a topological cell remain in this generalized framework. Thus we obtain example classes for the study of polytopes that are so fundamental in optimization.

4.2 Chirotopes, orientations of all simplices

When we define oriented matroids as chirotopes, they should be viewed as an abstract way of looking at orientations of all abstract or formal simplices. We can also say that we have an abstract sign of determinant function that does not violate any Grassmann–Plücker relation. We characterize oriented matroids as *chirotopes* as follows.

Definition 4.3 *[Chirotope] For a finite set $E = \{1, \ldots, n\}$ and*

$$\Lambda(n, r) := \{(\lambda_1, \ldots, \lambda_r) \in E^r \mid 1 \leq \lambda_1 < \cdots < \lambda_r \leq n\}$$

let $\chi : \Lambda(n, r) \to \{-1, 0, +1\}$ be a map, extended onto all r-tuples out of n elements according to the determinant rules (alternating and anti-symmetric). We require that $\chi^{-1}\{-1, 1\}$ forms an (ordinary) matroid, compare with Page 13. Furthermore, each restriction of χ onto $r + 2$ of the n elements can be obtained as a chirotope of some matrix with $r + 2$ rows. Then the map χ forms a chirotope with n elements in rank d.

The required existence of matrices in the definition for any subset of $r + 2$ elements does not imply the existence of a global matrix M such that the chirotope can be obtained as the signs of determinants of submatrices of M. However, this condition of the definition is equivalent to the condition that 3-term Grassmann–Plücker relations are not violated by χ, compare with Page 44. The latter condition can be checked easily by some Haskell code.

The name chirotope was suggested by Dreiding when he worked with Dress and Haegi on the classification of molecules, see Dreiding, Dress, and Haegi, 1982. It has turned out that the corresponding approach of using Grassmann–Plücker relations was studied several times before. Gutierrez Novoa seems to have been the first in 1965, see Gutierrez Novoa, 1965.

There is a recent new proof that shows the equivalence of oriented matroids defined as sphere systems with oriented matroids definded as chirotopes. The proof is similar to the rank 3 case of the previous chapter although the topological concepts in higher dimensions are more involved. We have decided not to include the proof here, see Bokowski, King, Mock, and Streinu, 2005.

4.3 Contraction, deletion, relabeling, reorientation

We have several operations on the set of oriented matroids that we have encountered already in the rank 3 case. We repeat these concepts here for sphere systems and chirotopes for arbitrary rank. A *sign reversal* is just a change of all signs of the chirotope or a way to orientate the main sphere of the sphere system.

We do not really change the oriented matroid by a sign reversal. A *relabeling* is just a renumbering of the elements that does not change the object itself. Very often we consider only the equivalence class up to relabeling the elements. We have to discard oriented matroids that are equivalent up to relabeling when we try to enumerate those equivalence classes. *Reorientating* an element changes the orientation of the chirotope for those simplices that contain this element. In the sphere system model we change the orientation of the corresponding pseudo-sphere. Again, many properties do not change when we reorientate some elements. This is why we often study the equivalence class of oriented matroids up to sign reversal, up to relabeling, and up to reorientation. In the sphere system model we can think of having no labels, no orientations for the pseudospheres, and no orientation for the global sphere. There is of course the corresponding equivalence class on the chirotope level. However, we work better with representatives for such a class. We recall our Haskell functions that were already written for the general rank case.

- The function `relabelChi::(Int->Int)->[OB]->[OB]` on Page 87 changes a chirotope according to a given permutation of its elements.
- A corresponding function that does the sign reversal is trivial.
- The function `reorSetChi::[Int]->[OB]->[OB]` from Page 86 does the reorientation of a chirotope for the elements of a given list of elements.
- The function `delSetChi::[Int]->[OB]->[OB]` from Page 22 deletes the elements of a given list of the chirotope. Deletion in the sphere system model simply means discarding corresponding pseudospheres.
- The function that does the contraction of a chirotope is more difficult to understand. In the sphere system model, we consider one pseudosphere as the new global sphere. The intersections of the remaining pseudospheres with this sphere form the new sphere system, the contraction at that pseudosphere. For chirotopes we have to think about a repeated abstract projection. Our function `ctrSetChi::[Int]->[OB]->[OB]` from Page 23 was already written for the general rank case.

4.3.1 *Mutations of a uniform chirotope*

Let us work with uniform chirotopes and the concept of a mutation to deepen the understanding of the chirotope axioms. The uniform case of a chirotope occurs if all orientations of simplices are different from zero. In this case the matroid property within the chirotope axioms is fulfilled. The underlying matroid is trivial. In order to confirm the chirotope property, we have to guarantee that there is no contradiction to any 3-term Grassmann–Plücker relation. We work with the following example of rank 4 with ten elements. We know that the signs in this example form a chirotope.

Chirotope example with 10 elements in rank 4

```
ex::[Int]
ex=[ 1,  1,  1,  1,  1,  1,  1,-1,-1,-1,-1,-1,-1,-1,-1,-1,-1,-1,  1,-1,  1,
    -1,-1,-1,-1,  1,-1,-1,-1,-1,-1,-1,-1,-1,-1,-1,-1,-1,-1,  1,-1,  1,
    -1,-1,-1,-1,  1,-1,-1,  1,  1,-1,  1,-1,-1,-1,-1,-1,-1,-1,-1,  1,-1,
    -1,  1,-1,  1,-1,-1,-1,-1,  1,-1,-1,  1,  1,  1,-1,  1,  1,-1,-1,-1,  1,
    -1,-1,-1,-1,-1,-1,  1,  1,  1,  1,-1,  1,-1,-1,-1,-1,-1,-1,  1,-1,-1,
    -1,-1,-1,-1,-1,-1,-1,  1,-1,-1,  1,-1,  1,-1,-1,-1,  1,-1,  1,  1,  1,
     1,-1,  1,  1,  1,-1,  1,  1,-1,-1,  1,-1,-1,-1,-1,-1,-1,-1,-1,-1,-1,
     1,-1,-1,  1,-1,  1,-1,-1,-1,  1,-1,  1,  1,  1,  1,-1,  1,  1,-1,  1,
     1,-1,-1,  1,-1,-1,-1,  1,  1,-1,  1,  1,-1,  1,-1,  1,  1,-1,  1,-1,  1,
    -1,-1,  1,-1,-1,-1,  1,-1,  1,-1,  1,-1,  1,-1,-1,-1,-1,  1,  1,  1,-1]
```

We discuss some Haskell code in which checking the chirotope condition leads to all mutations of a given chirotope.

Definition 4.4 *[Mutation of a uniform oriented matroid] Mutation of a uniform oriented matroid is a base that keeps the chirotope property if its sign changes.*

If we change the sign of a base, we only have to check those Grassmann–Plücker conditions in which this base occurs. The Grassmann–Plücker relation is used in the essential rank 2 case. We use a contraction argument for that. We construct a test chirotope by changing the sign of the given base. For the given base we construct all 3-term Grassmann–Plücker relations where this base occurs. The remaining details should be clear from the following Haskell program. Recall the Grassmann–Plücker relation structure and assume $a < b < c < d$.

$$\{u,v \,|\, a,b,c,d\} => (a\ b)(c\ d) - (a\ c)(b\ d) + (a\ d)(b\ c) = 0$$
$$ s1 \quad\ \ s2 \quad\quad s3 \quad s4 \quad\quad s5 \quad s6$$

How to find mutations of a chirotope in rank 4

```
gpCheckR2::[Int]->[OB]->Bool
gpCheckR2 [a,b,c,d] chi
 | s1*s2== -s3*s4 && s1*s2==s5*s6 = False
 | otherwise                      = True
 where
 s1=si a b chi;  s2=si c d chi
 s3=si a c chi;  s4=si b d chi
 s5=si a d chi;  s6=si b c chi


si::Int->Int->[([Int],Int)]->Int
si _ _ [] = 2
si a b (([c,d],s):hs) |[a,b]==[c,d]= s
                      |otherwise  = si a b hs
```

```
mutOfChi::[Int]->[OB]->Bool
mutOfChi [m,n,o,p] chi
 =and[gpCheckR2(sort[a,b,c,d])(ctrSetChi[u,v]testChi)
      |[u,v] <- tuplesL 2 [m,n,o,p],
       [a,b] <- tuplesL 2 ([m,n,o,p]\\[u,v]),
       [c,d] <- tuplesL 2 ([1..10]\\[u,v,a,b])]
 where
 testChi = [changeSign [m,n,o,p] el| el<-chi]
 changeSign::[Int]->OB->OB
 changeSign mut sTuple
  | mut == fst sTuple = (mut, -(snd sTuple))
  | otherwise         = sTuple
```

We form questionable chirotopes `testChi` by changing the sign of just one base. We consider all rank 2 contractions in which this base sign might lead to a contradiction. If there is no such contradiction, we have a mutation. The function `gpCheckR2::[Int]->[OB]->Bool` does the checking. We obtain 16 mutations as follows.

```
chirotope = zip (tuples 4 10) ex final= [el|el<-tuples 4 10,
mutOfChi el chirotope == True]

       [[1,2,3,4],[1,2,3,10],[1,2,5, 7],[1,4,5, 8],
        [1,7,8,9],[1,7,9,10],[2,3,5, 8],[2,3,5,10],
        [2,3,6,9],[2,3,8, 9],[2,6,8,10],[3,4,6, 7],
        [4,6,7,8],[4,7,8,10],[5,6,8, 9],[5,6,9,10]]
```

4.4 A polytope without diagonals need not be a simplex

We apply the chirotope concept for showing that there are convex polytopes in higher dimensions different from a simplex that have no diagonal. In dimension three the only convex polytope with this property is a tetrahedron, however, see Császár's polyhedron on Page 218 and neighbourly pinched spheres on Page 237. In linear optimization we walk from vertex to vertex along an edge so as to increase the objective function. Any pivot rule tells us how to proceed. A polytope without diagonals would allow us to reach the optimal vertex in one step. When we test pivot rules for all convex polytopes, we should think of the surprising fact that there are polytopes without diagonals.

We use cyclic polytopes for our assertion. The formal definition of a cyclic polytope will be given in the next section. We need here just a straightforward property of cyclic polytopes following directly from their definition: the chirotope is our alternating oriented matroid from Page 15.

Simplicial facets of a uniform chirotope

```
simplicialFacetsChi::[OB]->[[Int]]
simplicialFacetsChi chi
=[p|p<-tuples (r-1) n,
     (and[norm(p++[x], 1)'elem'chi|x<-[1..n]\\p])
   ||(and[norm(p++[x],-1)'elem'chi|x<-[1..n]\\p])]
 where r = length (fst (head chi))
       n = maximum (concat (map fst chi))
```

We have chosen the alternating oriented matroid with eight elements in rank 5 as an example. Although there exists a matrix for this oriented matroid, we confirm the chirotope property as an exercise on an abstract level. All simplex orientations are positive. The chirotope is uniform and the underlying matroid is trivial. Because of the special sign structure of the chirotope, inserting the signs in all 3-term Grassmann–Plücker relations leads to the same pattern. We have indeed a chirotope.

We now apply some Haskell code for our example chirotope. The list of facets will show us that there are no diagonals. Any $(r-1)$-tuple defines a hyperplane p. The condition for a facet is easy to formulate. The function `simplicialFacetsChi` returns all (simplicial) facets of the chirotope. We have to use only the functions `tuplesL` and `tuples` from Page 13, the function `norm` from Page 21, and the function `altOM` from Page 15.

A pair of vertices that lies in a simplicial facet is connected by an edge and cannot form the end points of a diagonal of the polytope. How many diagonals are left? We obtain all pairs that form no end points of diagonals with `map (tuplesL 2) (simplicialFacetsChi(altOM 5 8))`.

After concatenation, discarding double elements and sorting all pairs, we see that our alternating oriented matroid has no diagonal at all. After the hugs prompt we type `final` and the answer is `True` showing our assertion.

Final code for the *no diagonal*-property

```
res= map (tuplesL 2) (simplicialFacetsChi(altOM 5 8))
final= sort(nub(concat res))==tuples 2 8
```

Confirming the *polytope without diagonals* property exemplifies the chirotope setting in connection with some Haskell code.

4.5 Example classes of oriented matroids

We next provide some example classes of oriented matroids in higher rank.

Cubes and their chirotopes

```
cube2MA::Int->MA
cube2MA 1 = [[0],[1]]
cube2MA d =   (map([0]++)(cube2MA(d-1)))
            ++(map([1]++)(cube2MA(d-1)))
cube2MAHom::Int->MA
cube2MAHom d = map([1]++)(cube2MA d)


subMA::[Int]->MA->MA
subMA t m = map(i->m!!(i-1))t


cube2Chi::Int->[OB]
cube2Chi d
 =zip tu (map( ->toInt(signum(det
         (subMA t (cube2MAHom d)))))) tu )
 where tu = tuples (d+1) (2^d)
```

The chirotope of a *d*-dimensional cube can easily be generated inductively. The function `cube2Chi` determines the chirotope of a *d*-cube.

4.5.1 Cyclic polytopes

Oriented matroids of cyclic polytopes (=alternating oriented matroids) are even simpler than those of cubes. Understanding these polytopes is interesting in its own right.

Consider k points on the *moment curve*: $t \in \mathbb{R} \mapsto v(t) = (t, t^2, t^3, t^4) \in \mathbb{R}^4$, $P = \{v(t_1), v(t_2), \ldots, v(t_k)\}$, with parameters $t_1 < t_2 < \cdots < t_k$, $k \geq 6$. They define a *cyclic polytope* Z as their convex hull, $Z := conv\ P$. The point set P, written in homogeneous coordinates, yields

$$
M = \begin{pmatrix}
1 & t_1 & t_1^2 & t_1^3 & t_1^4 \\
1 & t_2 & t_2^2 & t_2^3 & t_2^4 \\
\vdots & \vdots & \vdots & \vdots & \vdots \\
1 & t_k & t_k^2 & t_k^3 & t_k^4
\end{pmatrix}
\begin{matrix}
\leftarrow j_1 \\
\\
\leftarrow \vdots \\
\leftarrow j_5
\end{matrix}
\tag{4.1}
$$

The determinant of a 5-row submatrix T of M corresponding to the indices $j_1 < j_2 < j_3 < j_4 < j_5$ can be evaluated using the Vandermonde determinant known from linear algebra or, for example, by studying interpolation polynomials of Lagrange.

$$\det T = \begin{vmatrix} 1 & t_{j_1} & \cdots & t_{j_1}^4 \\ 1 & t_{j_2} & \cdots & t_{j_2}^4 \\ 1 & t_{j_3} & \cdots & t_{j_3}^4 \\ 1 & t_{j_4} & \cdots & t_{j_4}^4 \\ 1 & t_{j_5} & \cdots & t_{j_5}^4 \end{vmatrix} = \prod_{\substack{(a,b) \\ 1 \le a < b \le 5}} (t_{j_b} - t_{j_a}) > 0 \qquad (4.2)$$

We consider hyperplanes H_{p_1,p_2,p_3,p_4}, $p_1 < p_2 < p_3 < p_4$, defined as affine hull of four points $v(p_1)$, $v(p_2)$, $v(p_3)$, $v(p_4)$ on the moment curve:

$$H_{p_1,p_2,p_3,p_4} = \{(w, x, y, z))| \begin{vmatrix} 1 & t_{p_1} & t_{p_1}^2 & t_{p_1}^3 & t_{p_1}^4 \\ 1 & t_{p_2} & t_{p_2}^2 & t_{p_2}^3 & t_{p_2}^4 \\ 1 & t_{p_3} & t_{p_3}^2 & t_{p_3}^3 & t_{p_3}^4 \\ 1 & t_{p_4} & t_{p_4}^2 & t_{p_4}^3 & t_{p_4}^4 \\ 1 & w & x & y & z \end{vmatrix} = 0\} \qquad (4.3)$$

Now the following is easy to confirm and we refer the reader to books about convex polytopes for a more detailed investigation. The key argument for the facet property of an affine hull of four vertices is to observe on which side of it we find all the remaining vertices, see, for example, Grünbaum, 2003.

1. H_{p_1,p_2,p_3,p_4} does not contain an additional point of the moment curve.
2. In the points $v(t_{p_1})$, $v(t_{p_2})$, $v(t_{p_3})$, $v(t_{p_4})$ the moment curve "intersects".
3. We find supporting hyperplanes of Z via **Gale's eveness condition**.
4. Each line segment joining two vertices of Z is an edge of Z.

The last assertion tells us that cyclic polytopes in dimension 4 have no diagonals, that is, line segments connecting two vertices that contain interior points of the polytope. We have explicitly confirmed this property in one example in the last section.

From the investigation of the cyclic polytope, we see how the chirotope can be obtained. It is given by a sequence of positive determinant signs, the most simple input structure for the omawin program (compare with Section 1.9).

4.5.2 Cyclic polytope in the plane

The cyclic polytope with seven elements in the plane leads to a rank 3 oriented matroid, its Folkman–Lawrence representation has been depicted in Figure 4.1.

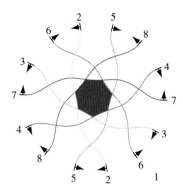

Figure 4.1 Folkman–Lawrence representation of a cyclic polytope in the plane

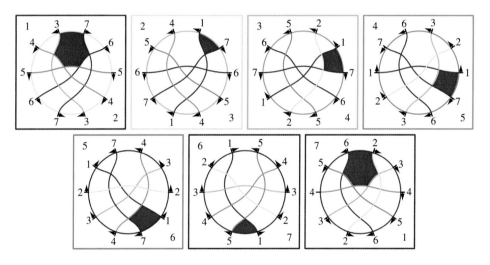

Figure 4.2 Folkman–Lawrence representation of a cyclic polytope in 3-space

4.5.3 Cyclic polytope in 3-space

The cyclic polytope with seven vertices in Euclidean 3-space leads to a rank 4 oriented matroid. Its Folkman–Lawrence representation has been depicted in Figure 4.2. We call a cell of maximal dimension in the Folkman–Lawrence representation a *tope*. The tope can be described by a sign vector. The tope corresponding to $(+, +, +, +, +, +, +)$ or $(-, -, -, -, -, -, -)$ has been indicated by its boundary elements. It is a wedge having a seven-gon as its top and its bottom. Note the cyclic structure, there are seven such wedge cells in a cyclic order. Adjacent topes have a hexagon in common.

4.5.4   Cyclic polytope in 4-space

A cyclic polytope in dimension 4 with seven vertices can be obtained as the convex hull of the following seven row vectors.

$$\begin{pmatrix} -3 & 9 & -27 & 81 \\ -2 & 4 & -8 & 16 \\ -1 & 1 & -1 & 1 \\ 0 & 0 & 0 & 0 \\ 1 & 1 & 1 & 1 \\ 2 & 4 & 8 & 16 \\ 3 & 9 & 27 & 81 \end{pmatrix}$$

A planar, dynamic Cinderella picture of a two-dimensional projection of these points can be used to show that all possible line segments are edges. It is easy to move four image points on four axes to appropriate points such that a given edge appears as a supporting edge of the projection. In this manner we can find all possible projections in the plane. The projection in Figure 4.3 shows for example

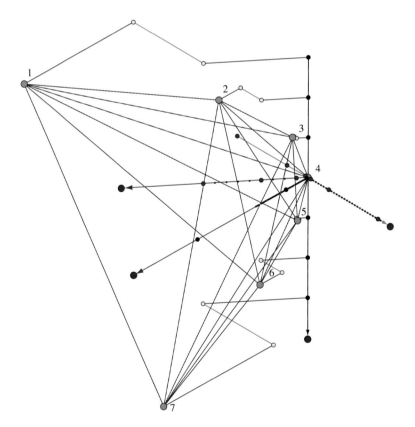

Figure 4.3   Projection of a cyclic polytope from 4-space

that the line segment from i to $i+1$ (modulo 7) for $i \in \{1, 2, \ldots, 7\}$ forms an edge of the polytope. Although the argument for that can be given without a picture, a student might like to play with the projections. A Cinderella picture can be changed with a mouse at the URL address of the author, `http://juergen.bokowski.de`. The corresponding Cinderella file can also be downloaded.

4.6 Hyperline sequences, algorithmic advantage

Before providing axioms for oriented matroids according to the hyperline sequences, we explain the underlying geometric idea of that concept in arbitrary dimensions.

We consider a vector arrangement $V = \{v_1, \ldots, v_n\} \subset \mathbb{R}^r$ of unit vectors that span $\mathbb{R}^r$, with $r \geq 2$. Let $B \subset \mathbb{R}^r$ be an oriented subspace of codimension 2 spanned by $V \cap B$. We obtain a vector arrangement $V_B = V \cap B$ in $\mathbb{R}^{r-2}$. The orthogonal complement C of B is a plane, that is orientated according to the orientation of B and of $\mathbb{R}^r$. The orthogonal projection of $V \backslash V_B$ to C is an ordered set of non-zero vectors which give rise to an ordered set L_B of oriented lines in $\mathbb{R}^2$. When we move along a circle in C around the origin according to the orientation of C, we meet the elements of V_C in a circular sequence Z_B, where any element of V_C is met twice (in positive and negative orientation). By an inductive definition, the *hyperline sequence of rank r* associated to B is the pair $(Y_B | Z_B)$, where Y_B is the oriented matroid of rank $r - 2$ associated to V_B. The oriented matroid of rank r associated to V is the set of all hyperline sequences that can be read off from V.

Let $(E, <)$ be a finite totally ordered set. Let $\overline{E} = \{\overline{e} | e \in E\}$ be a copy of E. The set $\mathbf{E}$ of *signed indices* is defined as the disjoint union of E and $\overline{E}$. By extending the map $e \mapsto \overline{e}$ to $\overline{e} \mapsto \overline{\overline{e}} = e$ for $e \in E$, we get an involution on $\mathbf{E}$. We define $e^* = \overline{e}^* = e$. For $X \subset \mathbf{E}$, define $\overline{X} = \{\overline{x} | x \in X\}$ and $X^* = \{x^* | x \in X\}$.

An *oriented d-simplex* in E is a $(d+1)$-tuple $\sigma = [x_1, \ldots, x_{d+1}]$ of elements of $\mathbf{E}$ such that $x_1^*, \ldots, x_{d+1}^*$ are pairwise distinct. Let an equivalence relation $\sim$ on oriented d-simplices in E be generated by $[x_1, \ldots, x_{d+1}] \sim [x_1, \ldots, x_{i-1}, \overline{x_{i+1}}, x_i, x_{i+2}, \ldots, x_{d+1}]$, for $i = 1, \ldots, d$. As usual, any oriented d-simplex is equivalent to one of the form $[e_1, \ldots, e_{d+1}]$ or $[e_1, \ldots, e_d, \overline{e_{d+1}}]$, with elements $e_1 < e_2 < \cdots < e_{d+1}$ of E. Define $-[x_1, \ldots, x_{d+1}] = [x_1, \ldots, x_d, \overline{x_{d+1}}]$. In the following inductive definition of hyperline sequences and oriented matroids, we denote with $C_m = (\{0, 1, \ldots, m-1\}, +)$ the cyclic group of order m.

Definition 4.5 *[Rank 1 oriented matroid] An oriented matroid X over $E(X) \subset E$ of rank 1 is a non-empty subset $X \subset E(X) \cup \overline{E(X)}$ such that $|X| = |X^*|$ and $X^* = E(X)$.*

The oriented simplex $[x]$ is by definition a *positively oriented base* of X for any $x \in X$. We define $-X = \overline{X}$.

Definition 4.6 *[Rank 2 oriented matroid] Let $k \in \mathbb{N}$, $k \geq 2$. A hyperline sequence X of rank 2 over $E(X) \subset E$ is a map from C_{2k} to oriented matroids of rank 1, $a \mapsto X^a$, such that $X^{a+k} = -X^a$ for all $a \in C_{2k}$, and $E(X) \cup \overline{E(X)}$ is a disjoint union of $X^0, \ldots, X^{2k-1}$.*

An oriented matroid of rank 2 is by definition a hyperline sequence of rank 2. We refer to $X^0, \ldots, X^{2k-1}$ as the *atoms* of X and to $2k$ as the *period length* of X. We say that $e \in E(X)$ is *incident* to an atom X^a of X if $e \in (X^a)^*$. Let $x_1, x_2 \in E(X) \cup \overline{E(X)}$ such that x_1^* and x_2^* are not incident to a single atom of X, and X induces the cyclic order $(x_1, x_2, \overline{x_1}, \overline{x_2})$. Then, the oriented simplex $[x_1, x_2]$ is by definition a *positively oriented base* of X. We define the hyperline sequence $-X$ over $E(-X) = E(X)$ of rank 2 as the map $a \mapsto (-X)^a = X^{-a}$ for $a \in C_{2k}$.

A hyperline sequence X of rank 2 is determined by the sequence $(X^0, \ldots, X^{2k-1})$ of atoms. We define that two hyperline sequences X_1 and X_2 of rank 2 are equal, $X_1 = X_2$, if $E(X_1) = E(X_2)$, the number $2k$ of atoms coincides, and X_1 is obtained from X_2 by a shift, that is, there is an $s \in C_{2k}$ with $X_1^{a+s} = X_2^a$ for all $a \in C_{2k}$.

We prepare the axioms for oriented matroids of rank $r > 2$ with the following definitions. A *hyperline sequence* X of rank r is a pair $(Y|Z)$, where Y is an oriented matroid of rank $r-2$ and Z is a hyperline sequence of rank 2. If X is a set of hyperline sequences of rank r, a *positively oriented base* of X in $(Y|Z) \in X$ is an oriented simplex $[x_1, \ldots, x_r]$ in $E(X)$, where $[x_1, \ldots x_{r-2}]$ is a positively oriented base of Y and $[x_{r-1}, x_r]$ is a positively oriented base of Z. Then, $-[x_1, \ldots, x_r]$ is a negatively oriented base of X. We define $-X = \{(Y|-Z) \mid (Y|Z) \in X\}$. An *atom* of X in a hyperline sequence $(Y|Z) \in X$ is the pair $(Y|Z^a)$, where Z^a is an atom of Z.

Definition 4.7 *[Oriented matroid of rank $r > 2$] A set X of hyperline sequences of rank r is an oriented matroid of rank $r > 2$ over $E(X) \subset E$ if it satisfies the following axioms.*

(H1) $E(X)$ *is a disjoint union of $E(Y)$ and $E(Z)$, for all $(Y|Z) \in X$.*

(H2) *Let $(Y_1|Z_1), (Y_2|Z_2) \in X$ and let $[x_1, \ldots x_{r-2}]$ be a positively oriented base of Y_1. If $\{x_1^*, \ldots, x_{r-2}^*\} \subset E(Y_2)$ then $(Y_1|Z_1) = (Y_2|Z_2)$ or $(Y_1|Z_1) = (-Y_2|-Z_2)$.*

(H3) *For all positively oriented bases $[x_1, \ldots, x_r]$ and $[y_1, \ldots, y_r]$ of X, there is some $j \in \{1, \ldots, r\}$ such that $[x_1, \ldots, x_{r-1}, y_j]$ is a positively or negatively oriented base of X.*

(H4) *For any positively oriented base $[x_1, \ldots, x_r]$ of X, $[x_1, \ldots, x_{r-3}, \overline{x_{r-1}}, x_{r-2}, x_r]$ is a positively oriented base of X.*

What do these axioms express? Let $V = \{v_1, \ldots, v_n\} \subset \mathbb{R}^r$, $B \subset \mathbb{R}^r$, Y_B, L_B, and Z_B be our set of vectors. To any $v_k \in V \backslash Y_B$ we get an oriented line $l_k \in L_B$. We move along a circle in the oriented plane and store the letter k in the circular sequence Z_B when l_k is met in positive orientation, and $\bar{k}$ if l_k is met in negative orientation. Obviously k and $\bar{k}$ appear on opposite places of the circular sequence. Hence Z_B is a hyperline sequence of rank 2. By induction and abuse of notation, the vector arrangement Y_B "is" an oriented matroid Y_B, and $(Y_B|Z_B)$ is a hyperline sequence of rank r. Axiom (H1) means that V is a disjoint union of $V \cap B$ and $V \backslash B$. Axiom (H2) corresponds to the fact that B is determined by any oriented base of V_B. Axiom (H3) is the Steinitz–McLane exchange lemma, stating that one can replace any vector in a base by some vector of any other base. Axiom (H4) ensures that the definition of oriented bases is compatible with the equivalence relation on oriented simplices. Axiom (H4) is related to the "consistent abstract sign of determinant" in Bokowski, Mock, and Streinu, 2001. It means that if r vectors span an $(r-1)$-simplex, then any subset of $r-2$ vectors spans a hyperline, and the orientation of the $(r-1)$-simplex does not depend on the hyperline on which we consider the r points.

We remark that the hyperline sequence concept is decisive for the extension algorithm in Chapter 7.

4.7 Circuits and cocircuits

We can find the definition of an oriented matroid by using circuit axioms of Folkman and Lawrence in the following form.

Definition 4.8 *An oriented matroid $\mathcal{O} = (E, \mathcal{C}, *)$ is a triple consisting of a ground set E, a fixed-point free involution $*$ on E and a collection of subsets $\mathcal{C}$ of E called circuits satisfying the axioms:*

(C1) $C, D \in \mathcal{C}$ implies that $C \not\subset D$.
(C2) $C \in \mathcal{C}$ implies that $C^* \in \mathcal{C}$ and $C \cap C^* = \emptyset$.
(C3) $C_1, C_2 \in \mathcal{C}$ $C_1 \neq C_2^*$, and $x \in C_1 \cap C_2^*$ implies that there is a $D \in \mathcal{C}$ so that $D \subset C_1 \cup C_2 \backslash \{x, x^*\}$.

Since showing axiom equivalence for oriented matroids is a difficult matter, we refrain from starting a corresponding discourse. We have seen in connection with oriented graphs that circuits and cocircuits are dual concepts of each other, compare with Page 65. The same holds true in the general case. We have worked out how to find the dual of a chirotope in Chapter 2.

Here we present some Haskell code that allows the computation of all cocircuits when a chirotope is given. The function `rankRm1` generates the list of all elements that lie in a rank $(r-1)$ submatroid. The function `chi2Coc` takes all of these lists, picks a basis within this submatroid and decides on which side the elements lie. When we think in linear algebra terms, we see that these ideas are a good guide to understanding the functions.

Cocircuits of a chirotope

```
chi2Coc::[OB]->[[Int]]
chi2Coc chi
 = sort(result++(map (map negate) result))
 where
  result = [ map (\i > sideOfHplane i hplane) [1..n]
             | hplane <- rankRm1 chi ]
  els   = nub(concat(map fst chi))
  r     = length (fst (head chi))
  n     = length els
  bases = [fst st| st <- chi, snd st 'elem' [-1,1]]
 sideOfHplane::Int->[Int]->Or
 sideOfHplane el hp
  | el 'elem' hp                           = 0
  | norm (((baseOfHp hp)++[el]),1)'elem' chi = 1
  | otherwise                              = -1
 baseOfHp::[Int]->[Int]
 baseOfHp hplane = head[el| el <- tuplesL (r-1) hplane,
                           x <- (els\\hplane),
                           sort (el++[x]) 'elem' bases ]
```

Flats of a chirotope of rank $r - 1$

```
rankRm1::[OB]->[[Int]]
rankRm1 chi=nub(map sort(nub[clHP hp chi|hp<-hPlanes]))
 where
  r     = length(fst (head chi))
  n     = maximum (concat (map fst chi))
  bases = [fst st| st <- chi, snd st 'elem' [-1,1]]
  hPlanes= nub[         fst (splitAt(j-1)base)
                   ++drop 1 ( snd (splitAt(j-1)base) )
                           | base<-bases, j <- [1..r]]
 clHP::[Int]->[OB]->[Int]
 clHP hp c = sort(hp++[el| el <- [1..n]\\hp,
                          (sort(hp++[el]),0)'elem'c])
```

We have shown, in the rank 3 case, some direct Haskell code that allows us to convert a chirotope into the list of all circuits. Since we obtain the circuits of the chirotope as the cocircuits of the dual chirotope, we have some code that also determines all circuits in the general case.

From the set of equivalence proofs for oriented matroid axioms, we select two versions in our next two theorems. They show that oriented matroids defined via chirotopes, via hyperline sequences, or via sphere systems coincide.

Theorem 4.9 *The set of positively oriented bases of a hyperline sequence of rank r over E is the set of positively oriented bases of a chirotope of rank r over E and vice versa.*

The rank 3 version equivalence is included in our Haskell program.

Theorem 4.10 *[A Topological Representation Theorem] To any hyperline sequence X of rank r over E_n, there is an arrangement $\mathcal{A}(X)$ of n oriented pseudo-hyperspheres in $\mathbb{S}^{r-1}$ of full rank with $X = X(\mathcal{A}(X))$. The equivalence class of $\mathcal{A}(X)$ is unique.*

We do not include the proofs here. The general case and an attempt to reduce the formalisms can be found in Bokowski, King, Mock, and Streinu, 2005. The proof uses induction on the number of elements and the rank of X. The names of Folkman and especially Lawrence have to be mentioned here.

We have mentioned several times that there are many more axiom systems for oriented matroids.

4.8   Extensions of oriented matroids

The inductive steps of our topological representation theorem indicate the algorithmic steps for finding all extensions of a given oriented matroid of fixed rank. The data structure of hyperline sequences are useful for this purpose.

There is a long history of programs that were developed in Darmstadt under the author's supervision or that the author wrote himself. In the beginning, that is, in the 1980s, the author's ideas led to a program in Fortran 77 implemented by Frederik Anheuser. Research results that emerged from that program often led either to coordinates or to non-realizability proofs. In these cases, the underlying program was at the end not decisive to a confirmation of the results. So it was perhaps only the Darmstadt school (with representatives like Sturmfels, Richter-Gebert, Guedes de Oliveira, Schuchert, and others) that knew about these tools.

Later there were two rank-independent programs written in the hope of improving the performance of the foregoing Fortran program, a C-program implemented by Guedes de Oliveira based on ideas same as those of hyperline sequences and a Pascal program implemented by Schuchert based on circuit axioms for oriented matroids. The performance was about the same in all cases. But these programs were used permanently to check corresponding results of all these programs. There was a resulting confidence that all these implementations lead to correct extensions. However, a bad performance behavior if applied for the challenging K_{12} problem in rank 4, was always unsatisfactory, compare Section 9.7. A minor but decisive idea of Bokowski and Guedes de Oliveira led to a breakthrough in this problem and related ones, see Bokowski and Guedes de Oliveira, 2000. Two additional independent extension programs which have used these improvements were written by Guedes de Oliveira in C and Bokowski in C++. An additional program, written in the functional language *Gofer*, served as an additional checking version. This implementation of Biermann, 1997, is available from the net under the URL address: `http:\\juergen.bokowski.de`. For polytope extensions we provide Haskell versions in rank 5 both for the uniform and the non-uniform cases of Chapter 7. We will discuss the details there.

The basic idea of determining all non-uniform extensions that Finschi used in his Ph.D. thesis, see Finschi, 2001, is closely related to this approach. Another such program was writtten by Klein from Cologne with supervision by Hochstättler, see Klein, 2002.

As a concluding remark we can say that all these attempts to implement an extension program for oriented matroids confirm its notable worth. So we will discuss such an algorithm in detail in the next section. There is also a recent program of D. Bremner for matroid polytope extensions and a Haskell version of Lars Schewe from Darmstadt. The C++-program of Bokowski for uniform rank 4 oriented matroid extensions can be downloaded from http://juergen.bokowski.de. It was used in Bokowski and Guedes de Oliveira, 2000 and by Frank Lutz.

5

From oriented matroids to face lattices

We have seen the transition from matrices to oriented matroids. If the matrix rows represent homogeneous coordinates of a point set, the combinatorial convex hull information is part of the oriented matroid information. This is one reason why oriented matroids are a useful concept for the theory of polytopes. Oriented matroids form a superset for studying combinatorial types of face lattices of convex polytopes "probably most of the convexity theory of compact subsets of R^n can be extended to the case of compact chirotopes, a program that might be worthwhile pursuing further," see Dress, 1986. An article entitled *Convexity in oriented matroids*, based on the circuit axioms, had already appeared, see Las Vergnas, 1980. Convex hulls of oriented matroids in rank 5 are interesting from a mathematical point of view. We therefore discuss combinatorial 3-spheres in particular.

By a polytope, we always mean a convex one, that is, the convex hull of a finite set of points in some finite Euclidean space. A j-face of a polytope is an intersection of a supporting hyperplane with the polytope P, the affine hull of which is j-dimensional. The set of all j-faces, $j = -1, 0, 1, \ldots, d$, of a polytope forms a lattice, its face lattice. The empty set with dimension -1 and the polytope itself with dimension d are called the improper faces.

Why do we study polytopes in higher dimensions?

In convexity, for example, in Brunn–Minkowski theory, see Schneider, 1993, we carry over results from polytopes via approximation techniques and continuity arguments to arbitrary convex bodies. Here polytopes serve as fundamental building blocks.

Every linear optimization problem deals with a polytope. A profound understanding of polytopes in high dimensions is mandatory in this mathematical field with its many applications. Moreover, in non-linear optimization problems, the linear case is often used repeatedly as an approximation tool.

Showing the beauty of polytopes and their historical development over decades, centuries, or even millennia, is another aspect. Handing over these mathematical concepts to posterity might be motivation enough for studying them.

The theory of polytopes has a long tradition. Among the latest books with many references to earlier work, we have *Lectures on Polytopes*, see Ziegler, 1994; the new edition of Grünbaum's book, *Convex Polytopes*, Grünbaum, 2003, and the book *Abstract Regular Polytopes*, see McMullen and Schulte, 2002.

The face lattice of a polytope occurs in the big face lattice of an oriented matroid

When we use the topological sphere system model for an oriented matroid, we have a topological cell decomposition of projective space. These cells of all dimensions form the *big face lattice* of an oriented matroid. In Figure 5.1 we see the big face lattice of a planar convex 30-gon. The Folkman–Lawrence representation of the rank 3 oriented matroid in which the 30 elements are the vertices again shows a 30-gon in which the elements are closed curves in the projective plane.

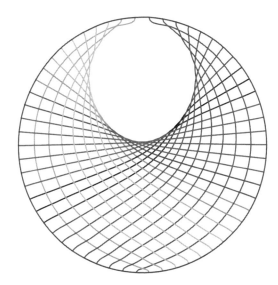

Figure 5.1   The surrounding cells of a convex 2-polytope within the big face lattice

When we start with the hyperplane arrangement of all supporting hyperplanes of a convex polytope, we also obtain such a cell decomposition, the big face lattice of the polytope. However, for polytopes the boundary structure, a combinatorial sphere, lies in the foreground and we do not work with the big face lattice very much. When we see the big face lattice of an oriented matroid as a generalization of a convex polytope, we understand the close interaction between the theory of oriented matroids and polytopes. Oriented matroid theory has not only stimulated research in the theory of polytopes but has provided additional tools. Combinatorial aspects of polytopes are better seen by taking an oriented matroid stance.

Spheres and face lattices of polytopes

For a d-polytope we obtain a combinatorial $(d-1)$-manifold from its combinatorial boundary structure, its face lattice. This combinatorial object, a combinatorial $(d-1)$-sphere, tells us many combinatorial properties of the given polytope. This mapping from the set of d-polytopes to the set of combinatorial $(d-1)$-spheres leads to a decisive problem when we look at the reverse problem. By starting with a $(d-1)$-manifold and considering whether or not it could be the boundary structure of a d-polytope, we are facing a hard question. We do not study general geometrical 3-manifolds. Our interest lies in the spherical case. We are often given a 3-manifold, perhaps knowing that it is a 3-sphere, and we want to decide whether this 3-manifold can arise as the boundary complex of a 4-polytope. So, it is useful to have some techniques for looking at spheres.

Polytopal and non-polytopal spheres

In a survey article, Mani-Levitska, 1993, we find this fundamental characterization problem of convex sets among five other characterization problems: *characterize, among all combinatorial $(d-1)$-spheres, the d-polytopal ones*. Because the general problem is considered to be wide open, our contribution can only be an interesting collection of examples.

Various topological methods have been applied and studied in order to cast new light onto the borderline between boundaries of 4-polytopes and 3-spheres. We mention King, 2001, and Richter-Gebert, 1996, the latter being a contribution to the realization spaces of polytopes.

Some interesting studies have involved investigating the finite world of spheres with given dimension and few vertices, see, for example, Altshuler, Bokowski, and

Steinberg, 1980, Mani-Levitska, 1972, and related papers. Another complementary area of research has been the asymptotical behavior of spheres. This topic was studied in Kalai, 1988, in which Kalai showed asymptotically that the polytopal spheres are just a very small subclass of combinatorial spheres. A corresponding result, see Pfeifle and Ziegler, 2002, deals with the asymptotical distribution of polytopal 3-spheres within the class of all combinatorial 3-spheres. In other words, the study of 3-spheres is an active area of research.

The question of how to characterize polytopal spheres among the class of combinatorial spheres has inspired much work in convexity. A local characterization of convex polytopes in higher dimension within the class of spheres has not been found. We do have corresponding satisfactory answers for 2-spheres due to Steinitz, 1922, see also Steinitz and Rademacher, 1934, however, the combinatorial 3-sphere case already turns out to be difficult to tackle. A recent problem class with contributions from Santos, Ziegler, and others has been considered in Paffenholz, 2004.

The content of this chapter

Our contributions in this chapter deal mainly with 3-spheres. Our discussion of face lattices of polytopes and more general face lattices, follows a set of examples. We start with a general convex hull implementation. It works in arbitrary dimensions and it uses the chirotope as its input. We determine the convex hull even in non-realizable cases.

We discuss the boundary structure of the four-dimensional cube. It explains a special Heegard splitting of a combinatorial 3-sphere. Next we discuss a face lattice that will turn out to be the boundary structure of a 4-polytope. We mention an interesting 3-sphere of Kleinschmidt. The combinatorial symmetry group of this sphere and the geometrical symmetry group of any realization of this sphere as the boundary of a polytope can never coincide. This result has led Mnëv to interesting investigations and striking results about realization spaces.

5.1   Convex hulls from chirotopes

Once we have the oriented matroid information, determining the convex hull is an easy task. From the oriented matroid information we compute the convex hull without using metrical concepts such as angles and distances.

Let us work on the following example where we have the chirotope of an 11-element point set in the plane. The matrix M contains 11 homogeneous coordinates and we compute the chirotope with our function `m2Chi` from Page 74.

$$M := \begin{pmatrix} 1 & 0 & 4 \\ 1 & 1 & 3 \\ 1 & 3 & 5 \\ 1 & 8 & 5 \\ 1 & 1 & 1 \\ 1 & 3 & 0 \\ 1 & 10 & 4 \\ 1 & 9 & 2 \\ 1 & 5 & 1 \\ 1 & 8 & 2 \\ 1 & 7 & 0 \end{pmatrix}.$$

Writing the full chirotope information with all tuples by hand is a boring task. However, it might be useful to have a look at this chirotope, so we provide this data in a table on Page 134.

For the chirotope information it is enough to know the second components `map snd result` as an ordered list according to the ordered tuples. We can use an additional function `change` to convert the integers into characters. This is a useful function for using the omawin program, compare with Page 299, since we need such input data.

Changing orientations $-1, 0, +1$ into signs $-, 0, +$.

```
change::[Int]->[Char]
change [] = []
change (x:xs)| x ==  0 = "0" ++ (change xs)
             | x ==  1 = "+" ++ (change xs)
             | x == -1 = "-" ++ (change xs)
```

We get the chirotope as the following sign list.

Chirotope as a sign list

```
> change (map snd result)
"++--++-++---------------++++++++-++-------++-----------
-----+++++++++-++-------++---------++++++++-++-------++-+
+++++++-++-------++-+++++-----+-+---+--++-------+++-++-"
```

We provide a convex hull algorithm implementation in any dimension that uses a chirotope as its input. The following description uses terms from Euclidean geometry, however, we think of these concepts in their abstract sense within the theory of oriented matroids.

Full chirotope information of matrix M

```
result = zip (tuples 3 11) chirotope
[([1, 2, 3], 1),([1, 2, 4], 1),([1, 2, 5],-1),([1, 2, 6],-1),([1, 2, 7], 1),
 ([1, 2, 8], 1),([1, 2, 9],-1),([1, 2,10], 1),([1, 2,11], 1),([1, 3, 4],-1),
 ([1, 3, 5],-1),([1, 3, 6],-1),([1, 3, 7],-1),([1, 3, 8],-1),([1, 3, 9],-1),
 ([1, 3,10],-1),([1, 3,11],-1),([1, 4, 5],-1),([1, 4, 6],-1),([1, 4, 7],-1),
 ([1, 4, 8],-1),([1, 4, 9],-1),([1, 4,10],-1),([1, 4,11],-1),([1, 5, 6], 1),
 ([1, 5, 7], 1),([1, 5, 8], 1),([1, 5, 9], 1),([1, 5,10], 1),([1, 5,11], 1),
 ([1, 6, 7], 1),([1, 6, 8], 1),([1, 6, 9],-1),([1, 6,10], 1),([1, 6,11], 1),
 ([1, 7, 8],-1),([1, 7, 9],-1),([1, 7,10],-1),([1, 7,11],-1),([1, 8, 9],-1),
 ([1, 8,10],-1),([1, 8,11],-1),([1, 9,10], 1),([1, 9,11], 1),([1,10,11],-1),
 ([2, 3, 4],-1),([2, 3, 5],-1),([2, 3, 6],-1),([2, 3, 7],-1),([2, 3, 8],-1),
 ([2, 3, 9],-1),([2, 3,10],-1),([2, 3,11],-1),([2, 4, 5],-1),([2, 4, 6],-1),
 ([2, 4, 7],-1),([2, 4, 8],-1),([2, 4, 9],-1),([2, 4,10],-1),([2, 4,11],-1),
 ([2, 5, 6], 1),([2, 5, 7], 1),([2, 5, 8], 1),([2, 5, 9], 1),([2, 5,10], 1),
 ([2, 5,11], 1),([2, 6, 7], 1),([2, 6, 8], 1),([2, 6, 9],-1),([2, 6,10], 1),
 ([2, 6,11], 1),([2, 7, 8],-1),([2, 7, 9],-1),([2, 7,10],-1),([2, 7,11],-1),
 ([2, 8, 9],-1),([2, 8,10],-1),([2, 8,11],-1),([2, 9,10], 1),([2, 9,11], 1),
 ([2,10,11],-1),([3, 4, 5],-1),([3, 4, 6],-1),([3, 4, 7],-1),([3, 4, 8],-1),
 ([3, 4, 9],-1),([3, 4,10],-1),([3, 4,11],-1),([3, 5, 6], 1),([3, 5, 7], 1),
 ([3, 5, 8], 1),([3, 5, 9], 1),([3, 5,10], 1),([3, 5,11], 1),([3, 6, 7], 1),
 ([3, 6, 8], 1),([3, 6, 9],-1),([3, 6,10], 1),([3, 6,11], 1),([3, 7, 8],-1),
 ([3, 7, 9],-1),([3, 7,10],-1),([3, 7,11],-1),([3, 8, 9],-1),([3, 8,10],-1),
 ([3, 8,11],-1),([3, 9,10], 1),([3, 9,11], 1),([3,10,11],-1),([4, 5, 6], 1),
 ([4, 5, 7], 1),([4, 5, 8], 1),([4, 5, 9], 1),([4, 5,10], 1),([4, 5,11], 1),
 ([4, 6, 7], 1),([4, 6, 8], 1),([4, 6, 9],-1),([4, 6,10], 1),([4, 6,11], 1),
 ([4, 7, 8],-1),([4, 7, 9],-1),([4, 7,10],-1),([4, 7,11],-1),([4, 8, 9],-1),
 ([4, 8,10],-1),([4, 8,11],-1),([4, 9,10], 1),([4, 9,11], 1),([4,10,11],-1),
 ([5, 6, 7], 1),([5, 6, 8], 1),([5, 6, 9], 1),([5, 6,10], 1),([5, 6,11], 1),
 ([5, 7, 8],-1),([5, 7, 9],-1),([5, 7,10],-1),([5, 7,11],-1),([5, 8, 9],-1),
 ([5, 8,10],-1),([5, 8,11],-1),([5, 9,10], 1),([5, 9,11],-1),([5,10,11],-1),
 ([6, 7, 8],-1),([6, 7, 9],-1),([6, 7,10],-1),([6, 7,11],-1),([6, 8, 9], 1),
 ([6, 8,10], 1),([6, 8,11],-1),([6, 9,10],-1),([6, 9,11],-1),([6,10,11],-1),
 ([7, 8, 9],-1),([7, 8,10],-1),([7, 8,11],-1),([7, 9,10], 1),([7, 9,11], 1),
 ([7,10,11], 1),([8, 9,10],-1),([8, 9,11], 1),([8,10,11], 1),([9,10,11],-1)]
```

The function `norm` from Page 21 returns an oriented simplex with unordered and perhaps negative elements in its normal form, that is, ordered, positive elements and a possible sign change according to the determinant rules. The function `rankRm1` from Page 126 starts with all non-degenerated simplices, discards one point at a time and uses the affine hull of the remaining points as a germ for determining all hyperplanes that are spanned by these subsets. This function can be used to determine all facets. If and only if there are points on both sides of the abstract hyperplane, we have no supporting hyperplane, that is, no facet set. These functions allow the convex hull of any point set in any dimension to be computed. Note that we are interested in having a method for determining the convex hull. An advantage of our approach is that we can form the convex hull of a chirotope on an abstract level, that is, independent of whether there are coordinates of the vertices. The convexity property of oriented matroids has been worked out by Las Vergnas. We can determine the convex hull in the abstract setting of oriented matroids, see Las Vergnas, 1980.

From the chirotope to its facets

```
chi2Facets::[OB]->[[Int]]
chi2Facets chi
 =[el| el <- rankRml chi,
    not(or[(norm(tu++[x], 1)'elem'chi
           &&norm(tu++[y],-1)'elem'chi)
         ||(norm(tu++[x],-1)'elem'chi
           &&norm(tu++[y], 1)'elem'chi)
             | tu  <-  tuplesL (r-1) el,
               tu 'elem' hps,
               [x,y] <- tuplesL 2 ([1..n]\\el) ])]
 where
 r = length(fst(head chi))
 n = maximum (concat (map fst chi))
 bases = [fst st| st <- chi, snd st 'elem' [-1,1]]
 hps   = nub[           fst(splitAt(j-1)base)
               ++drop 1 ( snd(splitAt(j-1)base) )
                 | base <- bases,  j <- [1..r] ]
```

We will test these functions later with the 3-cube example.

Convex hull in the uniform case is much simpler

When we know as in the case of a simplicial polytope that all extreme points of the convex hull lie in general position, we can use a much simpler method for finding the convex hull information.

The function `simplicialFacetsChi` checks for each hyperplane whether all remaining points lie on the same side and returns in this case the elements of the hyperplane. We have already used this function for the uniform case on Page 118.

The 3-cube as a non-uniform rank 4 example

We use a 3-cube as a rank 4 example. In homogeneous coordinates we get the matrix *M*. We obtain the chirotope as a result of the function `m2Chi4 8 cube3`.

Finally, with `chi2Facets` we get the six vertex sets of supporting hyperplanes of the 3-cube.

$$
M := \begin{pmatrix} 1 & 0 & 0 & 0 \\ 1 & 1 & 0 & 0 \\ 1 & 0 & 1 & 0 \\ 1 & 0 & 0 & 1 \\ 1 & 1 & 1 & 0 \\ 1 & 1 & 0 & 1 \\ 1 & 0 & 1 & 1 \\ 1 & 1 & 1 & 1 \end{pmatrix} \begin{matrix} 1 \\ 2 \\ 3 \\ 4 \\ 5 \\ 6 \\ 7 \\ 8 \end{matrix}.
$$

Homogeneous coordinates and facets of the 3-cube

```
cube3::[[Int]]
cube3 = [[1,0,0,0],[1,1,0,0],[1,0,1,0],[1,0,0,1],
         [1,1,1,0],[1,1,0,1],[1,0,1,1],[1,1,1,1]]
chi2Facets (m2Chi4 8 cube3)
   [[1,3,4,7],[1,2,4,6],[1,2,3,5],
    [2,5,6,8],[3,5,7,8],[4,6,7,8]]
```

5.2 A neighborly 3-sphere with ten vertices

Higher dimensional phenomena sometimes lead to unexpected results. We consider a polytope that has occured in Altshuler's investigation of neighborly 3-spheres. In Altshuler's sphere list of neighborly 3-spheres with 10 vertices it is No. 416. Let us confirm that the following coordinates of Bokowski provide us with a neighborly polytope, that is, that the polytope has no diagonals. Every pair of vertices of this polytope is connected along an edge of the polytope. Compare Bokowski and Sturmfels, 1987. We have studied a cyclic polytope with this property on Page 122.

We use our Haskell function `simplicialFacetsChi` to determine the facets. We obtain the list of simplicial facets below the set of coordinates.

This time we argue in a different way compared with Page 122. Denoting the number of j-dimensional faces of a d-dimensional convex polytope with f_j, we have Euler's formula in the form

$$
\sum_{j=1}^{d-1} (-1)^{j-1} f_j = 1 - (-1)^d.
$$

Coordinates and facets of Altshuler's sphere No. 416

```
m416::MA
m416=[[1,      0,      0,      0,      0],
      [1, 10000,      0,      0,      0],
      [1,      0, 10000,      0,      0],
      [1,      0,      0, 10000,      0],
      [1,      0,      0,      0, 10000],
      [1,   -132,    264,   9868,   1316],
      [1,  -1000,   3100,   7400,   1500],
      [1,  -3144,   9434,  -3144,   6289],
      [1,   2308,  -2564,   5128,  -5128],
      [1,  50000,  50000, -50000, -50000]]

[[1,2,5, 9],[1,2,5,10],[1,2,9,10],[1,3,8, 9],[1,3,8,10],
 [1,3,9,10],[1,4,5, 6],[1,4,5, 9],[1,4,6, 8],[1,4,7, 8],
 [1,4,7, 9],[1,5,6, 8],[1,5,8,10],[1,7,8, 9],[2,3,5, 7],
 [2,3,5, 8],[2,3,6, 7],[2,3,6,10],[2,3,8,10],[2,4,6, 9],
 [2,4,6,10],[2,4,9,10],[2,5,6, 7],[2,5,6, 9],[2,5,8,10],
 [3,4,7, 9],[3,4,7,10],[3,4,9,10],[3,5,7, 8],[3,6,7,10],
 [3,7,8, 9],[4,5,6, 9],[4,6,7, 8],[4,6,7,10],[5,6,7, 8]]
```

In our case we have $f_0 = 10$, $f_3 = 35$ and by counting the incidences of subfacets with facets, we find the special Dehn–Sommerville equation $2f_2 = f_3$. This confirms without checking the tetrahedra structure more closely: we have the maximal number of edges, namely $\binom{10}{2} = 45$. The polytope is neighborly.

For the Schlegel diagram notion see the next section. Determine the other facets of the dual 3-sphere and you will find an interesting fact about the symmetry of this example.

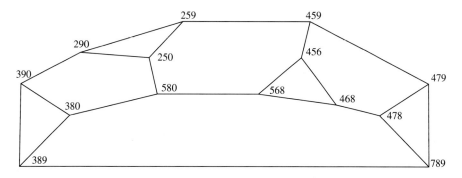

Figure 5.2   Schlegel diagram of a facet, dual 3-sphere of Altshuler's sphere No. 416

5.3   Heegard splitting of the 3-sphere

If mathematicians switch to higher dimensions and the spatial pictures vanish, not every student likes the subject any more. Therefore, we try to use pictures as often as possible to explain aspects of examples from four-dimensional space. The Schlegel diagram concept is a first concept in this direction.

Definition 5.1 *[Schlegel diagram] We obtain a Schlegel diagram of a polytope P if we use radial projection from a point, beyond a supporting hyperplane of a facet F of P and beneath all other supporting hyperplanes of facets of P, to project the polytope onto that facet.*

Figure 5.3 shows the edge skeleton of the 4-cube by parallel projection from 4-space. The right part shows the result of a radial projection, that is, a Schlegel diagram of the 4-cube. See also Figure 5.4 that shows a pottery model of the Schlegel diagram of the 4-cube.

The notion of a Schlegel diagram can be extended to an abstract Schlegel diagram notion on the level of oriented matroids. We use the chirotope model for an explanation. We can extend the chirotope by one additional element that lies *beyond a supporting hyperplane*, that is, we know the new signed base with respect to this cocircuit. In addition we choose the new element such that it "lies" *beneath all other supporting hyperplanes of facets*, that is, we know the signed base with respect to those cocircuits. The remaining signs can be chosen arbitrarily, however, such that we obtain a new chirotope. Contraction at this element leads to the Schlegel diagram notion for oriented matroids.

Face lattice of a 4-cube, seen in 3-space

For 4-dimensional polytopes the boundary is three-dimensional. The Schlegel diagram of a 4-polytope P was a first way to understand the face lattice of P in

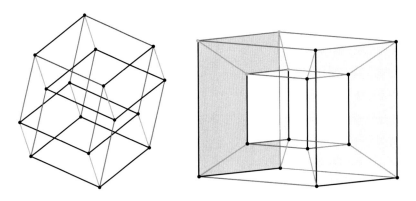

Figure 5.3   Edge skeleton of a 4-cube (left) and Schlegel diagram (right)

Figure 5.4   Pottery model of the Schlegel diagram of a 4-cube

3-space. We can describe the topological boundary structure of P in another way. We use for that the boundary structure of the 4-cube.

We see in Figure 5.5 a cut 3-cube and a cell subdivision of the boundary of the 4-cube as follows. The regular tetrahedron on the right, with suitable identifications along its boundary, forms a topological model of the topological

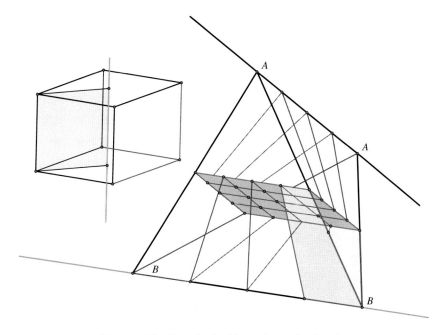

Figure 5.5   Topological boundary of a 4-cube

3-sphere. We identify the triangle on the right with the triangle on the left by rotating them around the A–A axis. In addition, we identify the front left triangle with the bottom one behind the tetrahedron by rotating it around the B–B axis. The reader can easily confirm that we have a three-dimensional manifold without boundary. Every point of the tetrahedron, including those of its suitable identified boundary, has a neighborhood that is homeomorphic to a three-dimensional ball. It is known that this 3-manifold is a 3-sphere. However, we can also argue that the eight cubes of the Schlegel diagram and the eight cubes in Figure 5.3 have the same incidence properties. Both face lattices are isomorphic. They form the boundary of the 4-cube which is a topological 3-sphere.

Bisecting the 3-sphere into two tori

We see in Figure 5.5 that the 3-sphere is subdivided into 8 cells each of which is combinatorially equivalent to a 3-cube. In other words, we see the subdivision of the boundary of the 4-cube. All 16 vertices of the 4-cube lie as lattice points on the square. The (4×4)-2 faces in this square form a 2-dimensional oriented manifold, a torus, which splits the 3-sphere into two solid tori. This is well known in topology and referred to as an example of a Heegard splitting of the 3-sphere into two tori. Note that by rotating the upper part of Figure 5.6 by 90 degrees, the corresponding 3-manifold obtained this way would not be a 3-sphere.

Another 3-sphere defined as a subdivided tetrahedron

The pottery model in Figure 5.6 shows in the same way the topological boundary of another combinatorial 3-sphere. It has six triangular prisms as facets. The

Figure 5.6　Topological boundary of a three-sphere with six triangular prisms, pottery model

3-cells are well defined in the two solid tori that correspond to their (identical) two-dimensional boundaries. We will present later some of our 3-sphere examples in this way, that is, by describing the 2-torus as the Heegard splitting of our 3-sphere.

How to find a 4-polytope with the 3-sphere as its boundary

Does there exist a 4-polytope whose face lattice coincides with the face lattice of the 3-sphere of Figure 5.6? We have $f_3 = 6$ facets that are triangular prisms. This implies a set of $f_2 = 15$ subfacets, and $f_0 = 9$ vertices appear on the inner square. Euler's formula in the four-dimensional case, $f_0 - f_1 + f_2 - f_3 = 0$, tells us the number of edges or 1-faces, $f_1 = 18$.

A 4-polytope corresponding to the polar dual 3-sphere would have each former vertex incident with 4 facets. We can label the vertices to obtain the following facets.

Simplicial 3-sphere of pottery model

```
facetsP::[[Int]]
 facetsP=[[1,2,4,5],[1,2,5,6],[1,2,4,6],
          [1,3,4,5],[1,3,5,6],[1,3,4,6],
          [2,3,4,5],[2,3,5,6],[2,3,4,6]]
```

With our methods, which will be explained for more complicated examples later, we can easily find coordinates for our polar dual pair of polytopes. The final chirotope information for the above example is the following:

```
{- [([1,2,4,5,6],1),([1,2,3,4,5], 1),([1,2,3,4,6],-1),
    ([1,2,3,5,6],1),([1,3,4,5,6],-1),([2,3,4,5,6], 1)]
```

Our approach for the study of 3-spheres is motivated by potential face lattices of 4-polytopes. Thus, the face-to-face cell subdivision of our objects plays a central role. If this subdivision does not lie in the foreground in topology, we are still interested in finding pictures or models that support our understanding. A nice model of Artmann in Figure 5.7 serves this purpose, it shows a Heegard splitting of the 3-sphere into two solid tori. The fibration on the boundary (Hopf fibres) indicates how the tori have to be glued, see also related figures in Section 7.8.

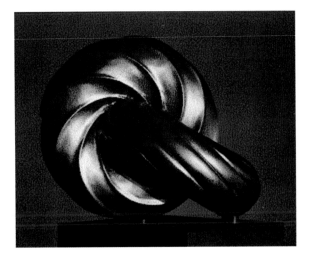

Figure 5.7   Two solid tori defining a 3-sphere; model of B. Artmann, Darmstadt

5.4   An interesting 3-sphere due to Kleinschmidt

The following matrix describes ten points in Euclidean 4-space given via its homogeneous coordinates. If we take the convex hull of these ten points, we obtain a 4-polytope. The oriented matroid information, a chirotope, can easily be computed.

$$
M := \begin{array}{c} 1 \\ 2 \\ 3 \\ 4 \\ 5 \\ 6 \\ 7 \\ 8 \\ 9 \\ 10 \end{array}
\left(
\begin{array}{rrrrr}
10000 & 0 & 0 & 0 & 0 \\
-2000 & -00259 & -1 & 400 & 800 \\
0 & 10000 & 0 & 0 & 0 \\
52636 & 5000 & 1 & -402 & -600 \\
0 & 0 & 1 & 0 & 0 \\
10000 & -10000 & -1 & 1000 & 1000 \\
0 & 0 & 0 & 1000 & 0 \\
0 & 0 & 0 & 0 & 1000 \\
-100000 & -9900 & -1 & 3000 & 38645 \\
-20000 & -1900 & 1 & 250 & 7695
\end{array}
\right).
$$

We have seen how we obtain the face lattice from the chirotope. If we use our Haskell code, we obtain the following simplicial face lattice. It is determined by all 28 of its 3-simplices.

This face lattice was found by Kleinschmidt who asked whether this 3-sphere can be realized as a polytope. Both Bokowski and Ewald gave an answer in the affirmative. Unfortunately, the coordinates in the corresponding paper Bokowski, Ewald, and Kleinschmidt, 1984, are not correct. However, correct coordinates were given later in Bokowski, 1991.

Facets of a 3-sphere of Kleinschmidt

```
facetsK::[[Int]]
facetsK=[[1,2,3, 4], [1,2,3, 7], [1,2,4, 8], [1,2,6, 7],
         [1,2,6, 8], [1,3,4, 7], [1,4,5, 6], [1,4,5, 8],
         [1,4,6, 7], [1,5,6, 8], [2,3,4, 8], [2,3,7,10],
         [2,3,8, 9], [2,3,9,10], [2,6,7, 9], [2,6,8, 9],
         [2,7,9,10], [3,4,5, 7], [3,4,5, 8], [3,5,7,10],
         [3,5,8,10], [3,8,9,10], [4,5,6, 7], [5,6,7, 9],
         [5,6,8,10], [5,6,9,10], [5,7,9,10], [6,8,9,10]]
```

As an application of our convex hull implementation, we can easily confirm that these coordinates are correct.

The interesting aspect of this 3-sphere concerns the symmetry of the face lattice. The combinatorial symmetry cannot be realized by an affine symmetry. This fact caused Mnëv to think about its realization space. see Mnëv, 1988, and more results about realization spaces in Richter-Gebert, 1996 and Richter-Gebert and Ziegler, 1995. Mnëv used this symmetry property of the 3-sphere and argued along Smith theory from topology that simplicial convex polytopes do not have the isotopy property. See also Bokowski and Guedes de Oliveira, 1990 and Paffenholz, 2004, for related results.

Mani-Levitska has shown that the relation between 3-polytopes and their combinatorial boundary structures, combinatorial 2-spheres, is different. Here the combinatorial symmetry of a 2-sphere can always be realized in 3-space with the same geometrical symmetry, see Mani, 1971. We later study other 3-spheres for which the combinatorial symmetry is different from their geometric symmetries for all realizations. A questionable, interesting 3-sphere of McMullen would have this property as well. However, here the realizability problem is still open, compare with Page 196.

6

From face lattices to oriented matroids I

For a given chirotope we have determined its face lattice of the convex hull on Page 133. In this chapter and the next we study the inverse problem. For a face lattice we search for an oriented matroid the convex hull of which has the given face lattice. We call such an oriented matroid a *matroid polytope*. Its existence is necessary for finding a convex polytope with the given face lattice.

How can we determine such a matroid polytope? Does this help to determine a corresponding convex polytope?

The answers are very easy in the two-dimensional case. Here we have just one polygon type for each number of vertices, namely that of a convex n-gon. In dimension 3 a characterization of face lattices of polytopes is also known. A theorem due to Steinitz characterizes all face lattices of 3-polytopes as planar graphs that are 3-connected. In other words, the graph can be embedded in the plane and it has at least four vertices any two of which can be joint by at least three independent paths, see Steinitz, 1922 and Steinitz and Rademacher, 1934. A profound discussion together with a new proof can be found in Ziegler, 1994.

Theorem 6.1 *[Steinitz] A graph is the edge graph of a three-dimensional convex polytope if and only if it is planar and 3-connected.*

The attempt to find a generalization of this theorem to higher dimensions has stimulated research on the borderline between potential face lattices of four-dimensional polytopes and questionable realizations. We emphasize the four-dimensional case by looking at some interesting examples in this chapter. We tackle the problem of finding for a given combinatorial 3-sphere a corresponding matroid polytope. Whereas a sphere of Brückner does not lead to a corresponding matroid polytope, Altshuler's sphere No. 963 does. We investigate this matroid polytope further and we determine a Heegard splitting. We discuss the dual sphere of Altshuler's sphere No. 963 and we show for this particular matroid polytope why it cannot occur as the boundary complex of a 4-polytope.

The whole chapter can be characterized as finding the matroid polytope directly. In the following chapter we discuss an algorithm that finds all matroid polytopes for given rank and given number of vertices. However, we emphasize the smallest interesting rank 5 case.

6.1   Finding matroid polytopes directly

Since there is no corresponding Steinitz theorem in higher dimensions, we have to find for each individual face lattice whether there exists a corresponding polytope. In many cases, we can effectively use the oriented matroid setting to answer that question. If there is no matroid polytope with the given face lattice, some programming can help to solve the question for small number of elements.

6.2   Two 3-spheres of Brückner and Altshuler

In this section we confirm the non-realizability of a 3-sphere with eight vertices due to Brückner, see Grünbaum, 2003, p. 222. We confirm that Altshuler's sphere M_{963}^9 with nine vertices does lead to a matroid polytope.

Haskell is again a good tool for our investigation. The functions `facetsB` and `facetsA` tell us the facets of these spheres.

Facets of a 3-sphere of Brückner

```
facetsB::[[Int]]
facetsB=[[1,2,3,4],[1,2,3,6],[1,2,4,6],[1,3,4,7],[1,3,6,8],
         [1,3,7,8],[1,4,5,6],[1,4,5,7],[1,5,6,8],[1,5,7,8],
         [2,3,4,8],[2,3,5,6],[2,3,5,8],[2,4,6,7],[2,4,7,8],
         [2,5,6,7],[2,5,7,8],[3,4,7,8],[3,5,6,8],[4,5,6,7]]
```

Faccts of a 3-sphere of Altshuler, No. 963

```
facetsA::[[Int]]
facetsA=[[1,2,5,6],[1,2,5,9],[1,2,6,9],[1,5,7,9],[2,6,8,9],
         [1,4,5,6],[2,3,5,6],[1,4,6,9],[2,3,5,9],[1,4,7,8],
         [2,3,7,8],[1,4,7,9],[2,3,8,9],[1,4,5,8],[2,3,6,7],
         [1,5,7,8],[2,6,7,8],[3,4,7,8],[3,4,7,9],[3,4,8,9],
         [3,5,6,7],[4,5,6,8],[3,5,7,9],[4,6,8,9],[5,6,7,8]]
```

For our reasoning, we assume that there is in all cases a polytope P with the given face lattice and we describe the case in which all vertices of P are in general position. This is in our simplicial case no loss of generality. We first look at all simplices with vertices from the polytope P that have two facets and a subfacet in common with P. For each subfacet of P we find precisely one such simplex. Without loss of generality we can fix the orientation of one of those simplices, say ([1,2,3,4,6],1) in the case of Brückner's sphere. When there is a polytope with the other simplex orientation, we can take a mirror image of that polytope. We call the affine hull of the subfacets, and more generally any affine space of codimension 2 spanned by vertices of P, a *hyperline*, compare with Page 47.

For an orthogonal projection of such a hyperline onto its orthogonal complement, see Figure 6.1. It shows the remaining projected vertices of P in a circular sequence from the image of the hyperline, compare Figure 2.2 on Page 47.

In the case of a simplicial polytope, we can assume that the vertices lie in general position. This implies that the projected vertices define a strict ordering for each hyperline. This circular sequence contains the information about the orientation of each simplex having a subfacet within the hyperline. In the case of a hyperline spanned by a subfacet [u,v,w], the two facets [u,v,w,x] and [u,v,w,y] meeting in the subfacet form *bounds* x and y within the circular sequence. We have used in this case the term *hyperline bounds* for the data structure [([u,v,w],[x,y]).

We obtain a subfacet of a simplicial polytope by deleting a vertex of a facet. We obtain all subfacets when we use the following Haskell function subfacets.

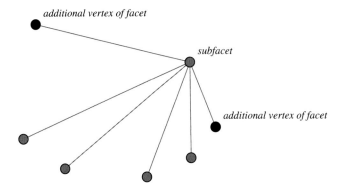

Figure 6.1   Hyperline bounds, projection along subfacet onto orthogonal plane

Subfacets of facets of a 3-sphere

```
subfacets::[[Int]]->[[Int]]
subfacets facets = nub (sub facets)
  where sub::[[Int]]->[[Int]]
        sub [] = []
        sub (f:fs) = [ f\\[j]| j<-f] ++ sub fs
```

We form all hyperline bounds with the function `hlbds`.

Hyperline bounds for all subfacets

```
hlBds::[[Int]]->[([Int],[Int])]
hlBds facets=[(sf,pair)|sf  <- subfacets facets,
                      pair <-
        let pairs = tuplesL 2 ([1..n]\\sf)
        in  pairs ++ (map reverse pairs),
             sort(sf++[head pair])'elem'facets,
             sort(sf++[last pair])'elem'facets]
    where n = length (nub (concat facets))
```

We find necessary simplex orientations via the function `nSO` when we compare a supporting hyperplane with all vertices that do not lie on this hyperplane.

Necessary simplex orientations

```
nSO::[[Int]]->[([Int],[Int])]->[OB]->[OB]
nSO _ [] so = so
nSO facets ( hlb@(hl,[x,y]):hlbRest) so
  |norm (hl++[x,y],1) 'elem' so
  = nSO facets [el|el<-hlbRest,el/=(hl,[y,x])]
    (so++[norm((hl++[x,i]),1)|i<-[1..n]\\(hl++[x,y])]
       ++[norm((hl++[i,y]),1)|i<-[1..n]\\(hl++[x,y])])
  |otherwise = nSO facets (hlbRest++[hlb]) so
     where n = length (nub (concat facets))
```

We use this function `nSO` to determine all new simplex orientations that we can get in this way, and we discard double elements. The intermediate result in both sphere cases is a list of signed bases.

Intermediate result

```
result1A=nub(nSO facetsA(hlBds facetsA)[(([1,2,5,6,9],1)])
result1B=nub(nSO facetsB(hlBds facetsB)[(([1,2,3,4,6],1)])
```

Now we use 3-term Grassmann–Plücker relations to find additional orientations of simplices. If five signs, orientations of simplices, within such a relation are known and the chirotope condition determines the unknown (because otherwise, we would have three equal monomial signs different from zero), we include that sign in our set of known ones. The function `consR5` on Page 148 does precisely that. The repeated application of `consR5` is done within the function `listOfNewSigns`. The function `ctrSetChi` of Page 23 reduces the oriented matroid to the rank 2 oriented matroid according to the projection along the hyperline.

Sign consequences from Grassmann–Plücker relations

```
consR5::[Int]->[Int]->[OB]->[OB]
consR5 hline [a,b,c,d] om2   -- new signs in rank 5
 |s3*s4/=0 && s3*s4+s5*s6==0
  if(s1==2&&s2/=0&&known 1)then=[norm(hline++[a,b],s2*s3*s4)]
  if(s2==2&&s1/=0&&known 2)then=[norm(hline++[c,d],s1*s3*s4)]
 |s1*s2/=0 && s1*s2==s5*s6
  if(s3==2&&s4/=0&&known 3)then=[norm(hline++[a,c],s4*s1*s2)]
  if(s4==2&&s3/=0&&known 4)then=[norm(hline++[b,d],s3*s1*s2)]
 |s1*s2/=0 && s1*s2+s3*s4==0
  if(s5==2&&s6/=0&&known 5)then=[norm(hline++[a,d],s6*s3*s4)]
  if(s6==2&&s5/=0&&known 6)then=[norm(hline++[b,c],s5*s3*s4)]
 |otherwise = []
 where
 s1=si a b om2; s2=si c d om2
 s3=si a c om2; s4=si b d om2
 s5=si a d om2; s6=si b c om2
 sv = [s1,s2,s3,s4,s5,s6]
known::Int->Bool
 known i = 2 'notElem' (take (i-1) sv ++ drop i sv)

si::Int->Int->[OB]->Or ; sign _ _ [] = 2
si a b (([c,d],s):hs) |(a,b)==(c,d)=s |otherwise = si a b hs
```

For the hyperline `hline` and four additional elements a, b, c, d, we have the following sign structure for the corresponding rank 2 contraction.

$$\{hline|[a,b,c,d]\}: \quad (a\,b)(c\,d) \; -(a\,c)(b\,d) \; +(a\,d)(b\,c)=0$$

$$s1s2s3s4s5s6$$

we use orientations $(-1,0,1,2)$, where $2 =$ sign unknown

We use the assumption a < b < c < d. Note that om is the contracted rank 2 oriented matroid whereas the resulting oriented base is a sign for the growing rank 5 chirotope.

We accumulate resulting signs according to this argument as the image of the function listOfNewSigns.

Resulting signs for a given partial chirotope

```
listOfNewSigns::Int->[OB]->[OB]
listOfNewSigns n oldSigns
 | newSigns==[] = sort oldSigns
 | otherwise    = listOfNewSigns n (oldSigns++newSigns)
 where newSigns = nub (concat set)
       set=[consR5 hline four (ctrSetChi hline oldSigns)
            | four <- tuplesL 4 [1..n],
              hline <- tuplesL (r-2) ([1..n]\\four)]
    r = length (fst (head oldSigns))
```

We obtain the resulting signs as result2B and result2A.

Resulting signs for our two spheres

```
result2A = nub (listOfNewSigns 9 result1A)
result2B = nub (listOfNewSigns 8 result1B)
```

It can happen that the chirotope condition is violated, that is, all non-zero monomials in a Grassmann–Plücker relation have equal sign. We check this via the functions contra and contraGP.

The function contraGP is decisive to confirming that Brückner's sphere cannot be polytopal. There is no compatible oriented matroid. The 3-sphere of Brückner allows not even a matroid polytope. However, there does exist a non-convex realization due to Mihalisin and Williams, 2002.

For Altshuler's sphere we obtain contraGP 9 resultA == True. All 126 signs have been determined, the chirotope that we have determined forms a matroid polytope. For the original result and a further result, see Altshuler, Bokowski, and Steinberg, 1980 and Bokowski and Schuchert, 1995. Our treatment is a method not only applicable for Brückner's or for Altshuler's sphere, but it also provides a general tool for spheres. Haskell was useful not only for providing the complete algorithm but also for its implementation.

Brückner's sphere is not polytopal, it leads to no matroid polytope

Altshuler's sphere does lead to a matroid polytope

```
contra::[Int]->[OB]->Bool
contra [a,b,c,d] list        -- a < b < c < d
  | s1*s2==  1 && s3*s4== -1 && s5*s6==  1 = False
  | s1*s2== -1 && s3*s4==  1 && s5*s6== -1 = False
  | otherwise = True
 where
 s1=si a b;s2=si c d;s3=si a c
 s4=si b d;s5=si a d;s6=si b c
 si::Int->Int->Int; si a b = sign a b list
contraGP::Int->[OB]->Bool
contraGP n oS =and[contra t4 (ctrSetChi t3 oS)
                 |t4<-tuples 4 n,t3<-tuplesL 3([1..n]\\t4)]
result3A = contraGP 9 result2A -- True, matroid polytope
result3B = contraGP 8 result2B -- False
```

We postpone the decision whether Altshuler's sphere No. 963 is polytopal. We first study Heegard splittings of it.

6.3   How to find a Heegard splitting of a sphere

Let us study Altshuler's simplicial complex from the last section once more as an example. We shall determine triangulated tori in the 2-skeleton of this complex. This investigation might be of interest in case of other given 3-manifolds. With additional arguments we can for example confirm the 3-sphere property this way, compare Figure 5.5 and remarks about bisecting the 3-sphere into two tori.

When we find a torus splitting or a more general genus g splitting of a 3-sphere that forms the boundary of a polytopal 3-sphere, we find a realization of this splitting 2-manifold without intersections in $\mathbb{R}^3$ within the Schlegel diagram of the polytope. Non-polytopal 3-spheres might lead to Heegard splitting 2-manifolds that are not embeddable without self-intersections, compare Page 253.

We provide some Haskell code for searching for orientable 2-manifolds in the 2-skeleton of 3-manifolds. We use all triangles in a 2-manifold that are incident with a fixed vertex, the *star* of that vertex. We assume that those edges of the star that are not incident with that vertex form a closed simple polygon, that is, the *link* of that vertex is a simple closed polygon.

We create the list of all subfacets and we form all possible stars with the function posStars.

Possible stars from a list of triangles

```
posStars::[[Int]]->[[Star]]
posStars triangles
  = map(\n->(nStars([t|t<-triangles,n'elem't]) n))
        (sort (nub (concat triangles)))

nStars::[[Int]]->Int->[Star]
nStars [] _ = []
nStars (t:ts) n
  =(completeStars (n,t\\[n]) ts)++(nStars ts n)
```

This is done by forming all possible stars for a fixed vertex number with the function nStars.

Here we use only triangles that contain vertex number *n*. The function nStars picks a first triangle from the list of triangles and forms with the function completeStars all stars with this triangle and adds to this list of stars the list of stars in which this triangle does not occur.

Complete a sub-star structure to all possible stars

```
completeStars::Star->[[Int]]->[Star]
completeStars (n,(x:xs)) trs
 |x==last xs = [(n,xs)]           -- completed link
 |trs == [] = []                  -- triangle list empty
 |otherwise
  =concat
    (map(\y->completeStars y [t|t<-trs, x'notElem't])
     [(n,(j:(x:xs)))|j<-(concat[t\\[n,x]|t<-trs,x'elem't])])
```

The function completeStars uses a growing star with center vertex *n* and a growing list that later will be the complete link. We assume that all the triangles for completing the star have the vertex *n*. When the last element of the growing link sequence is finally equal to the head element, we have a complete link and the star has been completed. When the remaining triangle list is empty before that happens, we can no longer form a star. The recursive structure of the function completeStars uses all triangles with vertices *j, x, n* and discards for later extensions all triangles with the already used vertex *x*.

For a compatible list of stars that forms a triangulated closed orientable 2-manifold, we can use Euler's formula to test whether we have a 2-sphere, a torus, or any other genus. The following three functions isSphere, isTorus, and genusManifold are straightforward applications of Euler's formula.

Is a given list of stars a 2-sphere or a torus?

```
isSphere::[(Int,[Int])]->Bool
isSphere stars = f_0 - f_1 + f_2 == 2
  where f_0 = length stars
        f_1 = div inciEdgeTriangle 2
        f_2 = div inciEdgeTriangle 3
        inciEdgeTriangle
           =sum(map length(map snd stars))
isTorus::[(Int,[Int])]->Bool
isTorus stars = f_0 - f_1 + f_2 == 0
  where f_0 = length stars
        f_1 = div inciEdgeTriangle 2
        f_2 = div inciEdgeTriangle 3
        inciEdgeTriangle
           =sum(map length(map snd stars))
```

From the list of stars of an orientable manifold to its genus

```
genusManifold::[(Int,[Int])]->Int
genusManifold stars = div (2 -(f_0 - f_1 + f_2)) 2
  where f_0 = length stars
        f_1 = div inciEdgeTriangle 2
        f_2 = div inciEdgeTriangle 3
        inciEdgeTriangle
           =sum(map length(map snd stars))
```

In order to find a list of consistent stars that can form an orientable 2-manifold, we use the function `adjEl` that determines the adjacent elements of a vertex *a* that lies in the link of another star. There is a condition for stars to be consistent, see Figure 6.2.

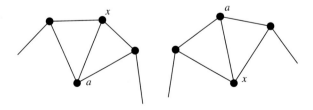

Figure 6.2 There is a condition for stars to be consistent. Adjacent triangles of an edge can occur in different stars

Find adjacent elements of *a* in link *ys* of a second star

```
adjEl::Star->Star->[Int]
adjEl (a,bs) (x,ys)  = [after a ys, before a ys]
   where after::(Eq a)=>a->[a]->a
         after p [] = p
         after p (u:us)
           |p==last us = u
           |p==u        = head us
           |otherwise  = after p us
         before::(Eq a)=>a->[a]->a
         before p ls = after p (reverse ls)
```

We have a consistent pair of stars when either the star indices do not occur in the link of the complementary stars or the former function `adjEl` leads to reversed pairs. This guarantees that we have used the same pair of triangles that is incident with the edge connecting the two star indices. This condition leads automatically to oriented manifolds.

We can use this function `consistentStarPair` repeatedly to check whether a star is consistent to an existing list of stars. This is done in the function `consistentStar`.

Is a star consistent to a list of stars?

```
consistentStar::Star->[Star]->Bool
consistentStar star partialManifold
  = and (map (\st->consistentStarPair star st)
    partialManifold)
```

```
consistentStarPair::Star->Star->Bool
consistentStarPair (a,bs) (x,ys)
  |(a 'notElem' ys) && (x 'notElem' bs)= True
  |(adjEl(a,bs)(x,ys))
     ==(reverse(adjEl(x,ys)(a,bs))) = True
  |otherwise                        = False
```

When we already have a list of stars, we can select those with a given vertex with the function `nstars`.

Select from a list of stars those with given vertex *x*

```
nstars::Int->[Star]->[Star]
nstars _ [] = []
nstars x starlist@(y:ys)
  |x==fst y  = [y]++nstars x ys
  |otherwise = nstars x ys
```

We form all orientable 2-manifolds in the 2-skeleton of the simplicial 3-sphere successively by adding stars consistently to all stars that have been chosen to be part of the 2-manifold.

The function `manifolds` uses a list of stars. For the first star we form all extensions to orientable manifolds. Then we drop this star and we apply the function `manifolds` to the remaining list of stars.

The function `posStar` is used in the extension process. It returns a test star, when it is consistent with all stars in a growing partial 2-manifold.

The function `extsmani` starts with a list of stars that is a potential partial manifold, a list of star indices for stars that have to be added, and a list of stars that have not so far been used for the partial manifold.

Finding all orientable manifolds from a list of stars

```
manifolds::[Star]->[[Star]]
manifolds [] = []
manifolds (star:stars)
  =(extsmani [star] (snd star) stars)++manifolds stars
```

Return star when consistent with starlist

```
posStar::Star->[Star]->[Star]
posStar star@(a,bs) partialManifold
  |consistentStar star partialManifold=[star]
  |consistentStar(a, reverse bs)partialManifold
                              =[(a,reverse bs)]
  |otherwise =[]
```

Extensions of partial manifolds

```
extsmani::[Star]->[Int]->[Star]->[[Star]]
extsmani starsSoFar []    _ = [starsSoFar]
extsmani starsSoFar (n:ns) starlist
 =concat
  (map(\i->extsmani (starsSoFar++[i])
         (nub((ns++(snd i))\\[fst y|y<-(starsSoFar++[i])]))
          starlist )  consnst )
 where
  nst = nstars n starlist
  consnst = concat(map(\st->posStar st starsSoFar) nst)
```

With `nst` we denote the sublist of stars with index *n*. Furthermore, we have with `consnst` the list of stars with index *n* that are consistent with the partial manifold. We add one of these consistent stars (*i*) at a time and we try to extend this new partial manifold again in the same manner. We now have one additional star: `starsSoFar++[i]`. We can use the old starlist because the index *n* will not occur again. However, it is possible that additional star indices have to be considered. The evaluation stops when the list of star indices has become the empty list.

The final result can now be evaluated. It turns out that there are many tori in the 2-skeleton of this 3-manifold. We depict one of them in Figure 6.3.

The Heegard splitting can also be studied via the incidences of facets with other facets. The 2-torus within the 2-skeleton has been marked, see Figure 6.4.

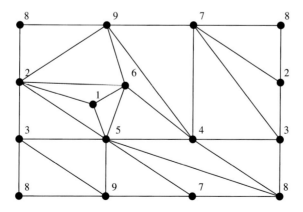

Figure 6.3   Torus within the 2-skeleton of Altshuler's sphere No. 963

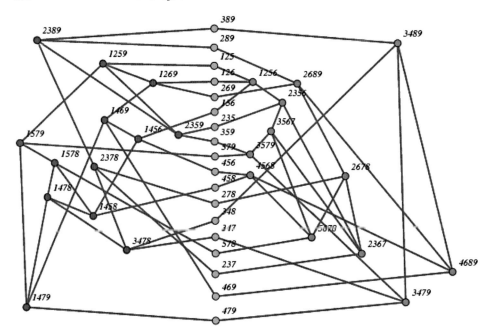

Figure 6.4   Incidence structure of simplices and torus triangles

Orientable triangulated 2-manifolds in the 2-skeleton

```
final = man (subfacets facetsA)

man::[[Int]]->[[Star]]
man triangles = manifolds(concat(posStars triangles))
```

The matroid polytope property provides us with the 3-sphere property as well. Our investigation was just an additional application of Haskell. However, we have learned something from our investigation about the symmetry of our simplicial complex, that is, about the maximal subgroup of the permutation group of these nine elements that keeps the simplicial complex fixed. By the way, all splitting tori of this sphere are realizable in $\mathbb{R}^3$.

6.4   The dual face lattice of Altshuler's 3-sphere No. 963

Now, we analyze the polar dual face lattice of Altshuler's 3-sphere No. 963. In this polar dual case the simplices become vertices, the old subfacets become edges, the edges become new subfacets, and the vertices become facets.

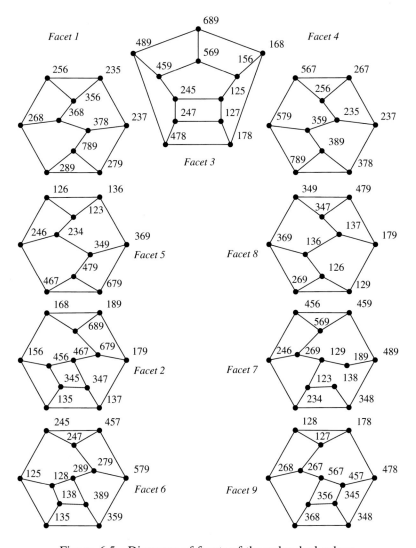

Figure 6.5   Diagrams of facets of the polar dual sphere

We depict these facets as diagrams homeomorphic to their Schlegel diagrams in Figure 6.5, again assuming that we can find them. The combinatorial types of these 3-cells admit a symmetry. A possible symmetry must keep facet 3 fixed but not necessarily its vertices.

The elements 1,4,5,8 and 2,6,7,9 can be permuted only among themselves. The cyclic group Z_4 acting on the set $\{2, 6, 7, 9\}$ can be discarded. Corresponding subfacets are not cyclically arranged. The only symmetry of facet 3, allowing an automorphism of the simplicial cell complex as well as that of its dual, is given by a group of order 2, Z_2, given by the permutation $\pi = (3)(1, 4)(2, 7)(5, 8)(6, 9)$.

6.5   Is Altshuler's 3-sphere No. 963 polytopal?

We assume that there is a polytope with the given boundary structure. The vertices 2,6,7,9 form a facet and vertex 3 does not lie in its affine hull. We can assume that in homogeneous coordinates our matrix has the form

$$
\begin{array}{c}
1 \\ 2 \\ 3 \\ 4 \\ 5 \\ 6 \\ 7 \\ 8 \\ 9
\end{array}
\left(
\begin{array}{ccccc}
. & . & a & c & b \\
1 & 0 & 0 & 0 & 0 \\
0 & 1 & 0 & 0 & 0 \\
h & . & e & . & f \\
l & . & i & k & j \\
0 & 0 & 1 & 0 & 0 \\
0 & 0 & 0 & 1 & 0 \\
p & . & m & o & n \\
0 & 0 & 0 & 0 & 1
\end{array}
\right)
$$

that is, the rows 2,3,6,7,9 form a unit matrix. We use the subfacet with vertices 2,7,9 first to find a partial information of the corresponding hyperline.

Hyperline subfacet	Facet Facet	Hyperline subfacet	Facet Facet
1 368	9 ...7.... 1		
2 136	8 ...2.... 5		
3 356	9....8.....1	359	4.....7....6
4 569	7..........3	567	4.....3....9
5 679	5....3.....2		
6 279	6....3.....1 start	129	8.....3....7
7 127	9...6,8....3		
8 237	4...6,9....1		
9 234	5....1.....7	347	2.....9....8

We consider 12 additional hyperlines defined by corresponding subfacets, assuming there is a polytope with facets of the given face lattice. Cutting a supporting hyperplane of a subfacet of the polytope at this affine hull of the subface (hyperline) and taking one half of it, if we turn that half hyperplane around the hyperline, we obtain a cyclic sequence of labels of all points that occur as incidences with that half-plane.

We start with subfacet 279 in line 6. Without loss of generality we pick the positive sign for the determinant $[23679] = +1$. The facet pairs have been chosen in such a way that the oriented simplices are consistent in the above partial scheme of all hyperlines. For example, facet 2679 defines for the above hyperline 5 whether the bounding elements have to be chosen in the given way or in reverse order.

In the same way, we can use facet 5679 to find orderings above. Facets 1356 and 1368 tell us how to choose the ordering in the first two hyperlines. Similarly, we get all the above orderings and we find the following values for ordered simplices

line	sign	value of determinant	No.
7	12379+	a > 0	(1)
7	12367-	-b > 0	(2)
9	23479-	-e > 0	(3)
9	23467+	f > 0	(4)
5	35679-	l > 0	(5)
1	36789+	-p > 0	(6)
9	12347-	af > be	(7)
9r	34789-	ep > hm	(8)
6r	12389-	cm > ao	(9)
3	35689-	lo > kp	(10)
2	12356-	bk > cj	(11)
4r	34567-	hj > fl	(12)
2	12368-	cn > bo	(13)
7	12378+	an > bm	(14)
9	23457+	fi > ej	(15)
3r	34579-	hi > el	(16)

We obtain a contradiction as follows.

```
(2),(9) and (1)(13) imply    cbm < abo < acn.
With (14) we have                c > 0   (17).
(3),(12) and (4),(16) imply  ehj < efl < fhi.
With (15) we have                h > 0   (18).
(5),(6),(7),(17) imply    acflp < bcelp.
(2),(5),(8),(17) imply    bcelp < bchlm.
(2),(5),(9),(18) imply    bchlm < abhlo.
(2),(5),(8),(17) imply    abhlo < abhkp.
(2),(5),(8),(17) imply    abhkp < achjp.
(2),(5),(8),(17) imply    achjp < acflp.
```

The last six inequalities imply the contradiction $acflp < acflp$. In other words, we have reached the conclusion that the given sphere is not polytopal.

Theorem 6.2 *[Bokowski] Altshuler's 3-sphere No. 963 cannot be realized as the boundary of a convex four-dimensional polytope.*

For the original proof see Altshuler, Bokowski, and Steinberg, 1980.

The smallest non-realizable substructure of that example still allows the same method of proof to show non-realizability. It has turned out that we can argue in the same way after contracting the oriented matroid at element 3. This was one way that a decisive oriented matroid with eight elements was born.

6.5.1   The study of 4-polytopes has an unexpected outcome

It is not only this example but the whole technique for tackling this problem that also has its applications in other areas. The model in Figure 6.6 and the pottery model in Figure 6.7 show the decisive rank 4 example as a sphere system. We come back to this example later when we study a mutation problem for oriented matroids.

Figure 6.6   Oriented matroid of rank 4, eight elements, seven mutations

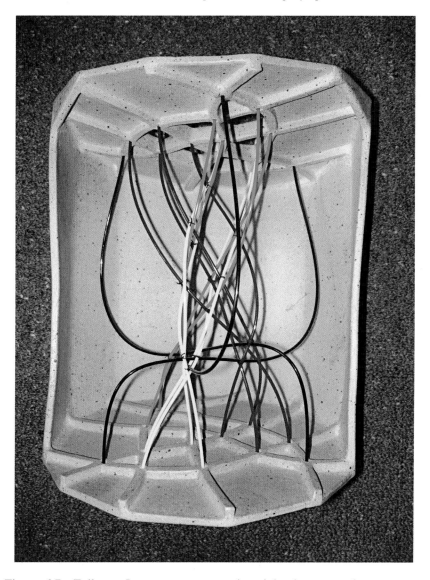

Figure 6.7 Folkman–Lawrence representation eight elements and seven mutations derived from Altshuler's sphere No. 963, pottery model

If we extend the signs of simplex orientations to a complete set of simplex
orientations, we find a Folkman–Lawrence representation of that corresponding
chirotope, that is, the 3-sphere can be viewed as the boundary of a topological
ball, the boundary objects being topological hyperplanes.

The signs for the chirotope of Altshuler's sphere were determined. We call a
sphere in this case *rigid*. The oriented matroid in terms of hyperline sequences is
the following.

Hyperline sequences for the matroid polytope of Altshuler's sphere No. 963

```
(123|4-5 9-6-7-8) (124|3 7 8 5-9 6) (125|3 9-6-4-8-7) (126|3 5-9-4-8-7)
(127|3-8-4 5-9 6) (128|3 7-4 5-9 6) (129|3 7 8 4 6-5) (134|2-7 9-8-5-6)
(135|2 6 4 8-9 7) (136|2 9-5 4 8 7) (137|2 8 4-5 9-6) (138|2-7 9 4-5-6)
(139|2-7 5-4-8-6) (145|2 6-8-7-3-9) (146|2 9-5-8-7-3) (147|2 6 5 8-9-3)
(148|2 6 5-7-3-9) (149|2 5 8 3 7-6) (156|2 9 3 7 8 4) (157|2 3 9-8-4-6)
(158|2 9 3 7-4-6) (159|2-7-3-8-4-6) (167|2 3-8-4 5-9) (168|2 3 7-4 5-9)
(169|2 5 3 7 8 4) (178|2 3 9 4-5-6) (179|2 3 5-4-8-6) (189|2 5-4 3 7-6)
(234|1 5 6 7-9 8) (235|1 9-6-7-8-4) (236|1 5-7-8-4-9) (237|1 5 6-8-4-9)
(238|1 5 6 7-9-4) (239|1 6 7 4 8-5) (245|1 9-6 3 7 8) (246|1 5 3 7-9 8)
(247|1-8 9 3-6-5) (248|1 7 3-6 9-5) (249|1-8 6-3-7-5) (256|1-3-7-8-4-9)
(257|1 4 8-3 6-9) (258|1 4-7-3 6-9) (259|1 6 4 8 7 3) (267|1 9 4 8-3-5)
(268|1 4 9-7-3-5) (269|1-8-4-7-3-5) (278|1 5 6-3-9-4) (279|1 6-3 4 8-5)
(289|1 4 6-3-7-5) (345|1-6-2 7-9 8) (346|1 5-2 7-9 8) (347|1 9-8-2-6-5)
(348|1 7-9-2-6-5) (349|1 5 6 2 8-7) (356|1 9 2-7-8-4) (357|1 4 8 2 6-9)
(358|1 4 9-7 2 6) (359|1 7-2-6-8-4) (367|1 9 4 8 2-5) (368|1 4 9-7 2-5)
(369|1 5-8-4-7 2) (378|1 5 6 2-4-9) (379|1 6 2 8 4-5) (389|1 5 6 2-4-7)
(456|1 9 2 3 7 8) (457|1-6-2-3-9 8) (458|1-6-2-9-3-7) (459|1-2-6 7 3 8)
(467|1 5-2-3-9 8) (468|1 5-9-2-3-7) (469|1 5 7 3 2 8) (478|1 5 6 2 9 3)
(479|1-3-8-2-6-5) (489|1 7 3-6-2-5) (567|1 9 2 3-8-4) (568|1 9 2 3 7-4)
(569|1-3-7-8-4 2) (578|1-6-2-3-9-4) (579|1 4 8 6 2 3) (589|1 4-3-7 6 2)
(678|1 5-2-3-9-4) (679|1-2-3 4 8-5) (689|1 4-2-3-7-5) (789|1 5 6 2-4 3)
```

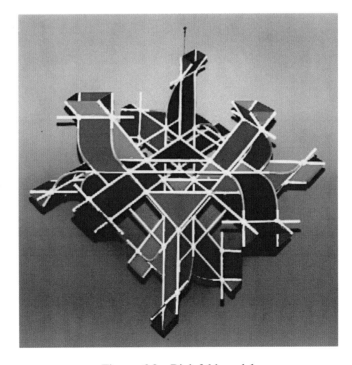

Figure 6.8   Bielefeld model

This matroid polytope was also used in Bokowski and Schuchert, 1995b, to show that is has no polar. This example is minimal with respect to its rank and its number of elements.

In general the chirotope is not uniquely determined as in the previous case of Altshuler's sphere or in the cases of neighborly polytopes, see Shemer, 1982.

The model of Figure 6.6 was exhibited in the City Hall of Bremen on the occasion of the 100th anniversary of the German Mathematical Society in 1990, see Figure 11.2. A later model was produced for the Centre of Interdisciplinary Research in Bielefeld, see Figure 6.8.

7

From face lattices to oriented matroids II

Within this chapter we describe an extension algorithm for matroid polytopes that has led to many decisions in various contexts. We present the uniform case first.

An application has led to an interesting matroid polytope with a Folkman–Lawrence representation of an equifacetted 3-sphere with ten Dürer polytope facets. We discuss this sphere with its interesting Heegard splitting in detail and we provide all rank 3 contractions of the matroid polytope. Similarly, Altshuler's sphere No. 425 leads to a matroid polytope and it has an interesting Folkman–Lawrence representation as well. The reasoning as to why these examples cannot occur as the boundary of a polytope can be very much condensed.

We finally deal with the extension algorithm in the non-uniform case that is more involved. A self-polar-dual sphere serves as an example to show the competition between a general algorithmical method of finding admissible oriented matroids for a given face lattice of a questionable polytope and intuitive guesses that might lead to a realization directly. For a 3-sphere of McMullen in connection with a dual pair of higher dimensional analogues of the Platonic solids in dimension 4, the problem of finding a matroid polytope is open. The same problem remains for a 3-sphere with 240 equal facets, pyramids over regular pentagons. However, it can be realized as a non-convex star shaped polyhedron by using precisely all vertices of the 600-cell.

7.1 Matroid polytopes via extensions

Assume that we have a partial piece of information for an oriented matroid, say a partial (possibly empty) list of its signed bases or oriented simplices. We provide an extension algorithm that determines all matroid polytopes that are compatible with the given information. We use the data structure of hyperline sequences

together with the chirotope setting to find these extensions. Consider the extension
by one element as the reverse operation of deleting one element. We proceed
step by step by inserting the new element in all possible positions and with all
admissible signs. During this process we check the sign condition of our growing
partial oriented matroids.

The uniform case of this extension algorithm is easier to implement. Corresponding programs of various performances have been used by the author and
his research students during the 1980s and 1990s. The results have often been
presented without referring to the extension algorithm if the oriented matroid has
finally led to coordinates. A rough early description has been given in Bokowski
and Sturmfels, 1989b.

7.2   Rank 5 uniform matroid polytopes

We describe within this section the uniform case. This case is much simpler
compared with the non-uniform case that we are going to describe in more detail
later. For finding simplicial 4-polytopes it suffices to understand the algorithm
in the uniform case. We explain the kernel part, the extension within one row of
the hyperline representation, the crucial part of the extension technique. We look
at the decisive function `ext` first. It determines for a sign s of the new element
n its position p, and for the uniform rank 2 contraction along the hyperline its
one element extension.

Let us study the function `inRow` in the uniform case. The variable `hyp`
represents the list of all hyperline sequences that we extend row by row, compare
for example the hyperline sequences on Page 162. The variable `chi` represents the
signs of all abstract simplices that we so far know. Splitting `hyp` at the position
`row` leads to the actual row data structure (`tr,om2`) as the head of the second
component of the function `splitAt`. We insert the new signed element $s \times n$ in
`om2` by using the function `ext` in all possible ways. The variable `st` stores the
list of new signs that we know after the insertion has been completed in this row.
We compare the new signed element $s \times n$ with all other elements in this row to
obtain new signs of abstract simplices.

When we cannot pick `newsigns`, we do not get an extension. This occurs when
we have a sign contradiction that will be detected in the function `newOrEmpty`.
This function `newOrEmpty` compares the preliminary sign list `chi` with `st`.

Within the list `tupels 5 finalN` we determine the position i of the actual
tuple `tu` and we find the corresponding sign `e`. When this sign `e` is different

from 2, that is, it has been determined, and when it is not equal to s, we obtain a contradiction, that is, the result is the empty set.

Thus the function inRow leads to a list of all extensions within the row under consideration together with new signs that are compatible with the given sign vector.

Insertion in a row in the uniform case

```
inRow::Int->Int->([[(Ngon,OM2)]],[Or])->[([[(Ngon,OM2)]],[Or])]
inRow n row (hyp,chi)
 = [(((aHyp++[(tr, ext s n p om2)]++bHyp), newsigns)
     | s<-[-1,1], p<-[1..lom2], newsigns<-
   let st= [norm(tr++[el,s*n],1)|i<-[1..p],    el<-om2!!(i-1)]
       ++[norm(tr++[s*n,el],1)|i<-[(p+1)..lom2],el<-om2!!(i-1)]
   in  newOrEmpty n chi st ]
 where (aHyp,((tr,om2):bHyp)) = splitAt (row-1) hyp
       lom2 = length om2
       ext::Int->Int->Int->OM2->OM2    -- s sign of n
       ext   s   n    p   om2 = a++[[s*n]]++b
        where  (a,b) = splitAt p om2
       newOrEmpty::Int->[Or]->[(Tu,Or)]->[[Or]]
       newOrEmpty n chi [] = [chi]
       newOrEmpty n chi ((tu,s):rest)
        |e'notElem'[s,2] = []
        |otherwise = newOrEmpty n newChi rest
        where  i  = head(elemIndices tu(tuples 5 finalN))
          (a,(e:b))= splitAt i chi; newChi= a++[s]++b
```

Now a frame is missing that uses the function inRow repeatedly and that does the next extension when we do not extend the matroid polytope by just one element.

We use the function Chi2HypR5 which can convert a chirotope in rank 5 into the data structure of hyperline sequences. We also use a function for the reverse, the transition from hyperline sequences to its chirotope: hyp2ChiR5. We call the main function inAll. It uses the function inRow that generates all insertions of the new element in one row that are compatible with the partial chirotope that has so far been determined. The function extMPu finally produces all compatible extensions up to the final number of elements.

We understand the non-uniform case better when the uniform part is completely clear. So make sure that you did understand the uniform case. In the next section we deal with the non-uniform case in detail. You can skip it when you are satisfied with the remark that this is a more involved part, however, it uses the same underlying idea. It determines all matroid polytopes in the non-uniform case.

Matroid polytope extension, frame functions

```
inAll::Int->Int->Int->([OM5],[Or])->[([OM5],[Or])]
inAll  finN  n    row        sofar
 |row == 1 =inRow n 1 sofar
 |otherwise=concat(map(inRow n row)
                  (inAll finN n(row-1)sofar))


extMPu::Int->Int->[Or]->[[Or]]
extMPu finN k knownSigns
 | k == 5 = map snd (inAll finN 6 10 (startHyp,knownSigns))
 |otherwise
  =concat
   (map(\sl->
   let slist=delSetChi[k..finN](zip(tuples 5 finN)sl)
   in  map snd(inAll finN k rows
         (chi2HypR5 slist, sl) ))
         previousSignlists)
 where
  startHyp           = chi2HypR5 [([1,2,3,4,5],1)]
  rows               = length (tuples 3 (k-1))
  previousSignlists  = extMPu finN (k-1) knownSigns

erg=extMPu 7 7 [1]++drop 1(replicate(length(tuples 5 7))2)
```

7.3 Rank 5 non-uniform matroid polytopes

We describe now along with some Haskell code the inductive generation of all rank 5 matroid polytopes in the non-uniform case.

The face lattices that we obtain in this way contain all of those of 4-polytopes. Because of the decisive role of this extension algorithm, we present this part independently of all other parts. Redundancy can help to deepen the understanding. The reader who has carefully studied the functions of previous sections can simply skip them. We have listed the pages on which those functions occured for the first time. The novice in functional programming should at least consult the sublist of basic Haskell functions that we have provided within the Haskell primer, see Appendix A.

7.4 Contraction for matroid polytopes

We use within our extension algorithm for matroid polytopes the contraction at a set of elements of an oriented matroid. We have copied the corresponding Haskell functions once more in order to see the complete algorithm. Compare the idea of a contraction at certain elements with the case where the elements are points in an affine space. We generate the affine hull of the given elements and we use a projection along this hull onto its orthogonal complement. When we use vectors that point from the image of the affine hull to the images of the remaining elements, we obtain a vector model of a new oriented matroid, the contraction of the original oriented matroid of the point configuration along the elements of the affine hull.

All r-tuples of a list of integers

```
tuples::Int->Int->[[Int]]
tuples 0 n = [[]]
tuples r n = tuplesL r [1..n]


tuplesL::Int->[Int]->[[Int]]
tuplesL   r list@(x:xs)
    | length list <  r = []
    | length list == r = [list]
    | r == 1           = [[el]|el<-list]
    | otherwise = [[x]++el| el<-tuplesL (r-1) xs]
                     ++tuplesL r xs
```

We first provide functions that we use in this context: `tuples`, `tuplesL`, `norm`, `delSetChi`, `delElChi`. The function `tuples` from Page 13 returns all r-tuples of the list of the first n natural numbers and the function `tuplesL` from the same page returns all r-tuples of any given list of integers. These functions occur so often that we assume that no further comment is needed.

We interpret the pair of a list of integers and a sign as an oriented (abstract) simplex. The function `norm` from Page 21 returns such an oriented simplex with positive and sorted elements whereby the sign has changed accordingly. Think of the rules for determinants. Again, this is a basic function that we do not even mark in the following dependency analysis of all functions that we use in the generating algorithm of all matroid polytopes, compare Page 174.

Deleting a set of elements in a chirotope simply means that we discard all oriented simplices that contain elements from the set. The `delSetChi` function from Page 22 does precisely that. The function `delElChi` does the same for a single element.

Normal representation of an oriented simplex

```
norm::OB->OB
norm (tu@(h:rest),s) = normPos (list,s*signum prod)
  where prod = product tu; list = map abs tu
        normPos::OB->OB
        normPos  (tuple@(h:rest),sign)
          |rest == [] = ([h],sign)
          |h==minimum tuple=([h]++fst next,snd next)
          |odd (length rest) =normPos (rest++[h],-sign)
          |otherwise         =normPos (rest++[h], sign)
          where  next = normPos (rest,sign)
```

Deletion of an element or of a set of elements in a chirotope

```
delSetChi::[Int]->[OB]->[OB]
delSetChi [] chi = chi
delSetChi (h:rest) chi = delElChi h (delSetChi rest chi)
delElChi::Int->[OB]->[OB]
delElChi el [] = []
delElChi el (h@(tu,_):rest)
  |el'elem'tu = delElChi el rest
  |otherwise  = [h]++delElChi el rest
```

As we have already mentioned, contracting a set of elements corresponds to an (abstract) projection along the (abstract) affine hull, see also Page 23. For contracting at a set, we first delete those elements that lie in the (matroidal) closure of the maximal independent set.

The function `findRankInd` from Page 23 determines for a set and a chirotope a pair consisting of the rank of the set within this chirotope and a maximal independent set for that set. We need this function within the function `ctrSetChi`.

Contraction at a set or at a single element of a chirotope

```
ctrSetChi::[Int]->[OB]->[OB]
ctrSetChi [] chi = chi
ctrSetChi set@(h:rest) chi
 |k == length set = ctrSetChi rest (ctrElChi h chi)
 |otherwise = ctrSetChi list (delSetChi (set\\list) chi)
 where (k,list)=findRankInd set chi
ctrElChi::Int->[OB]->[OB]; ctrElChi k [] = []
ctrElChi k (h:list) = ((ctrST k h)++ctrElChi k list)
 where ctrST::Int->OB->[OB]
       ctrST k (tu,sign)
        |k 'notElem' tu    = []
        |posk 'mod' 2 == 0 = [(tu\\[k],      sign)]
        |otherwise         = [(tu\\[k], -1*sign)]
        where posk = head (elemIndices k tu)
```

Finding the rank and a base of a flat in a chirotope

```
findRankInd::[Int]->[OB]->(Int,[Int])
findRankInd  set chi
 |or[length(base\\set)==r-k|base<-bases]= (k,set)
 |otherwise = maximum (map(ı->findRankInd
              (take(i-1)set++drop i set)chi)[1..k])
  where k    =length set
        r    =length(fst (head chi))
        bases=[fst st|st<-chi,snd st'elem'[-1,1]]
```

7.5 Chirotopes and hyperline sequences

For a hyperline sequence representation of a rank 5 matroid polytope we can use a special data structure. Our convexity requirement implies that we never have three elements within one line. Moreover, each two-dimensional affine hull of vertices of a convex polytope is convex again. This implies that we can assume that each hyperline is that of a planar n-gon. The corresponding rank 2 oriented matroid can be described via the cyclic order of these elements. We can store the rank 5 oriented matroid as a list of pairs of k-gons, with $k \geq 3$ depending on the hyperline, together with rank 2 oriented matroids, the hyperline sequences, that is, rank 2 contractions at these hyperlines. The latter has a circular structure and we can assume that they have the smallest element with positive sign within their first set. For the *n*-gons we can also assume that their lists begin with the smallest element of each *n*-gon.

From a hyperline sequence data structure of a matroid polytope, we obtain the chirotope information easily via the function `hyp2ChiR5`. For one row in the hyperline representation we obtain from `hyp2ChiR5` a 5-tuple with a zero sign either when two elements belong to the same list within the rank 2 oriented matroid or when we can choose four elements within the first component of that row. Compare this with the rank 3 case from Page 82.

Rank 5 chirotope from hyperline sequences

```
hyp2ChiR5::[(Tu,OM2)]->[(Tu,Or)]
hyp2ChiR5 []=[]
hyp2ChiR5  ((hl,om2):rest)
  =nub(sort
      [norm(tr++[b,c],1)| [i,j]<-tuples 2 l,
       b<-om2!!(i-1), c<-om2!!(j-1), tr<-tuplesL 3 hl]
    ++[ norm (tr++[b,c],0)| i<-[1..l],
       [b,c]<-tuplesL 2(om2!!(i-1)),tr<-tuplesL 3 hl]
    ++[ norm (qu++[x],0)  | qu<-tuplesL 4 hl,
                                x<-(hl++concat om2)\\qu]
    ++ hyp2ChiR5 rest) where  l = length om2
```

The reverse transition from the chirotope to its hyperline sequence representation is more difficult to understand. We first repeat the decisive rank 2 case from Page 74 with the function `chi2HypR2` from Page 79. We start with the list of all one element lists and we form the union of such lists whenever a zero orientation of a pair of elements calls for that. The function `genSets` generates all the lists that have to be glued in this manner. In the second step we insert the signs for all elements. This is again a two step process. The signs for all elements that do not belong to the first list can be inserted first, compare `signsTail`. Afterwards, we use the inserted signs as a reference for the signs in the first list, compare `signsHead`. The function `preHypR2` returns the list of lists with its inserted signs. Finally, we sort these lists according to the chirotope with a special sorting function `chi2ordf` and we obtain via `chi2HypR2` the rank 2 oriented matroid in its hyperline sequence form.

We form all elements `els` and its triples `trs` to find all hyperlines `hls` as matroidal closures of such triples. We use each hyperline only once according to the function `nub`. The function `hl2Pair` returns the pair of the ordered hyperline and the rank 2 oriented matroid. With `chi2HypR5` we can evaluate the hyperline sequence representation from the chirotope.

Changing a rank 2 chirotope to a hyperline

```
chi2HypR2::[(Tu,Or)]->OM2
chi2HypR2 chi=map nub(sortBy(chi2ordf chi)(preHypR2 chi))


chi2ordf::[(Tu,Or)]->([Int]->[Int]->Ordering)
chi2ordf chi
  = f where f::[Int]->[Int]->Ordering
            f a b |norm([head a, head b],1)'elem'chi = LT
                  |otherwise                         = GT
```

Hyperlines for rank 5 matroid polytopes

```
chi2HypR5::[(Tu,Or)]->[OM5]
chi2HypR5 chi = map (\hl->hl2Pair hl chi) hls
 where trs  = tuplesL 3 els
       hls  = nub(map sort (map (\tr-> cl tr chi) trs))
       els  = sort(nub(concat(map fst chi)))
cl::[Int]->[(Tu,Or)]->[Int]
cl tr chi = tr++[e| e<-els, eInHl e tr els chi]
 where els = sort(nub(concat(map fst chi)))
```

eInHl

```
eInHl::Int->Ngon->[Int]->[(Tu,Or)]->Bool
eInHl e tr [] chi = True
eInHl e tr (x:xs) chi
  |x 'elem' (tr++[e])                = eInHl e tr xs chi
  |norm(tr++[e,x],0)'notElem'chi = False
  |otherwise                     = eInHl e tr xs chi
```

We will repeatedly change the data representation during the inductive generation of all matroid polytopes. We do not aim to present a very fast generation
algorithm for matroid polytopes but a compact executable version.

preHypR2, matroidal information for `chi2HypR2`

```
preHypR2::[(Tu,Or)]->[[Int]]
preHypR2  chi = siIn (genSets chi) chi

siIn::[[Int]]->[(Tu,Or)]->[[Int]]
siIn preom2 chi=signsHead chi(signsTail chi preom2)
  where
    els=(concat preom2);min=minimum els;ord=chi2ordf chi
    signsTail::[(Tu,Or)]->[[Int]]->[[Int]]
    signsTail chi preom2@(l:ls)
     =[l]++[ map signT set | set<-ls]
     where
     signT::Int->Int
     signT el | ord [min] [el] ==LT = el |otherwise= -el
     where min = minimum l
    signsHead::[(Tu,Or)]->[[Int]]->[[Int]]
    signsHead chi preom2@(l:ls)
    =[[min]++map(\el->signH el chi) (l\\[min]) ]++ls
     where
     signH::Int->[(Tu,Or)]->Int
     signH el chi|ord[el][last(last preom2)]==LT =el
                 |otherwise = -el
```

genSets, first preparational step for `chi2HypR2`

```
genSets::[(Tu,Or)]->[[Int]]
genSets chi = sort(glue chi (tuplesL l els))
 where
 els=sort (nub (concat (map fst chi)))
 glue::[(Tu,Or)]->[[Int]]->[[Int]];    glue [] pre = pre
 glue (([u,v],s):xs) pre
  |s==0= glue xs([set|set<-pre,intersect set [u,v] == []]
    ++[nub(concat[a++b|a<-pre,b<-pre,u'elem'a,v'elem'b])])
  |otherwise = glue xs pre
```

`hl2Pair`, essential step to form the pair of a hyperline with its sequence

```
hl2Pair::Ngon->[(Tu,Or)]->OM5
hl2Pair hl chi=(nGonSort hl om2, om2)
 where
 om2 = chi2HypR2 (ctrSetChi hl chi)
 nGonSort::[Int]->OM2->[Int]
 nGonSort hl om2 = [m]++sortBy fSort (hl\\[m])
  where   m     = minimum hl
          [x,y]= [head(head om2),last(last om2)]
          fSort::Int->Int->Ordering
          fSort a b|norm ([m,a,b,x,y],1)'elem'chi = LT
                   |  otherwise                   = GT
```

7.6 Matroid polytope extensions, function dependencies

We see in Figure 7.1 a circle on the right with all functions that provide the contraction of a matroid polytope.

The circle in the middle contains the functions that determine the hyperline structure of a matroid polytope from its chirotope. As we know, a main part forms

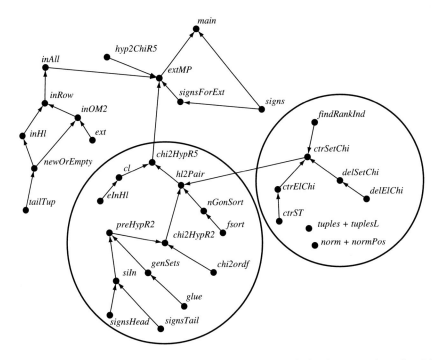

Figure 7.1   Haskell functions and their dependencies within the extension algorithm

the transition in the rank 2 case which has been discussed earlier. We look now at the actual and decisive extensions.

7.7 The essential part of the extension algorithm

We use as an example the generation of all matroid polytopes with seven elements. We start with a simplex in dimension four. The chirotope, as a list of oriented bases, is in this case just `[([1,2,3,4,5],1)]`.

```
startHyp =  ([[1,2,3],[[4],[ 5]]),    ([1,2,4],[[3],[-5]]),
            ([1,2,5],[[3],[ 4]]),    ([1,3,4],[[2],[ 5]]),
            ([1,3,5],[[2],[-4]]),    ([1,4,5],[[2],[ 3]]),
            ([2,3,4],[[1],[-5]]),    ([2,3,5],[[1],[ 4]]),
            ([2,4,5],[[1],[-3]]),    ([3,4,5],[[1],[ 2]])]

take 1 (extMatrPoly 7 7 (signs 7))
[([[([1,7,6,2,3],[[4],[5]]),
        ([1,2,4],[[3,-6,-7],[-5]]),
        ([1,2,5],[[3,-6,-7],[4]]),
        ([1,3,4],[[2,6,7],[5]]),
        ([1,3,5],[[2,6,7],[-4]]),
        ([1,4,5],[[2],[3],[-7],[-6]]),
        ([1,4,6],[[2,3,-7],[5]]),
        ([1,5,6],[[2,3,-7],[-4]]),
        ([2,3,4],[[1,6,7],[-5]]),
        ([2,3,5],[[1,6,7],[4]]),
        ([2,4,5],[[1],[7],[6],[-3]]),
        ([2,4,6],[[1,3,7],[-5]]),
        ([2,5,6],[[1,3,7],[4]]),
        ([3,4,5],[[1],[7],[6],[2]]),
        ([3,4,6],[[1,-2,7],[-5]]),
        ([3,5,6],[[1,-2,7],[4]]),
        ([4,5,6],[[1],[7],[-2],[-3]])],
     [0,0,0,-1,0,0,-1,0,0,-1,1,0,0,-1,-1])
```

We determine the corresponding hyperline sequences of this matroid polytope with the function `chi2HypR5` and we call this `startHyp` within the function `extMP`.

We think of doing a reverse deleting in this data structure. We work at the same time with the chirotope data structure.

Our function `extMP` uses as pre-image the final number of elements of the list of matroid polytope extensions and a chirotope as start matroid polytope.

extMP

```
main = print (map snd (extMP 7 7 (signs 7)))

extMP::Int->Int->[Or]->[([OM5],[Or])]
extMP n k signs
  | k == 6  = inAll 6 10 (startHyp, startSigns)
  |otherwise= concat ( map (inAll k lTu3s)
            [(el, signsForExt n k signs)
              |el<-map chi2HypR5(map hyp2ChiR5
               (map fst(extMP n (k-1) signs)))])
  where startHyp  = chi2HypR5 [([1,2,3,4,5],1)]
        startSigns= signsForExt n 6 signs
        lTu3s     = length (tuples 3 (k-1))
```

tailTup

```
tailTup::Int->[[Int]];tailTup n=map(++[n])(tuples 4 (n-1))

signsForExt::Int->Int->[Or]->[Or]
signsForExt n k signs
  =map (i-> signs!!(head (elemIndices i (tuples 5 n))))
       (map (++[k]) (tuples 4 (k-1)))
```

inAll

```
inAll::Int->Int->([OM5],[Or])->[([OM5],[Or])]
inAll n row sofar
  |row == 1  =inRow n 1 sofar
  |row>maxRow=concat(map(inRow n maxRow)
                        (inAll n(maxRow-1)sofar))
    |otherwise =concat(map(inRow n row)
                        (inAll n(row-1)sofar))
  where maxRow = length (fst sofar)
```

Our result of the function `inAll` will be a list of matroid polytope extensions. Each single matroid polytope is a pair consisting of its hyperline representation and its chirotope.

In our example we do not know any sign in advance, so we prepare `startSigns` with the function `signsForExt` that gives us a list of unknown signs (= 2).

inRow

```
inRow::Int->Int->([OM5],[Or])->[([OM5],[Or])]
inRow n row pair = (inHl n row pair)++(inOM2 n row pair)
inHl::Int->Int->([OM5],[Or])->[([OM5],[Or])]
inHl n row (rows,chi)
 =[(((firstRows++[(take g gon++[n]++drop g gon,om2)]
                              ++lastRows),signs)
   | g<-[1..length gon], signs <-
 let si= [norm(tr++[p1,p2],1)
            |tr<-tuplesL 3 (take g gon++[n]++drop g gon),
            n'elem'tr, [p1,p2]<-pairs om2]
        ++[norm(tr++[p1,p2],0)
            |tr <- tuplesL 3 (take g gon++[n]++drop g gon),
            n'elem'tr, u<-[1..length om2],
            [p1,p2]<-tuplesL 2(om2!!(u-1))]
        ++[norm(tr++[n,x],0) |tr <- tuplesL 3 gon,
                        u<-[1..length om2],
                        x<-(om2!!(u-1))++(gon\\tr)]
   in       newOrEmpty n chi si                        ]
 where (firstRows,((gon,om2):lastRows))=splitAt(row-1)rows
        pairs::OM2->OM2
        pairs om2=[[x,y]| [u,v] <- tuples 2 (length om2),
                        x<-om2!!(u-1), y<-om2!!(v-1)]
```

Rank 2 extension with signed element *n*

```
ext::Int->Int->Int->Int->OM2->OM2
ext s n p q om2|q==0=take (p-1) a++[(last a)++[s*n]]++b
               |q==1=a++[[s*n]]++b
    where(a,b)=splitAt p om2
> ext (-1) 6 2 0 [[1],[2],[3,4],[5]]
leads to [[1],[2,-6],[3,4],[5]]
```

newOrEmpty

```
newOrEmpty::Int->[Or]->[(Tu,Or)]->[[Or]]
newOrEmpty n chi [] = [chi]
newOrEmpty n chi ((tu,s):rest)
 |e'notElem'[s,2]= []
 |otherwise      = newOrEmpty n newChi rest
  where  i        = head (elemIndices tu (tailTup n))
         (a,(e:b))= splitAt i chi; newChi= a++[s]++b
```

inOM2

```
inOM2::Int->Int->([OM5],[Or])->[([OM5],[Or])]
inOM2 n row (rows,chi)
=[(((firstRows++[(gon,ext s n p q om2)]
              ++lastRows),newsigns)
            | s <- [-1, 1], p <- [1..lom2], q <- [0,1],
              newsigns<-
  let s1=if q==0
      then[norm(tr++[el,s*n],1)|i<-[1..p-1],el<-om2!!(i-1),
           tr<-trs ]++[norm(tr++[el,s*n],0)|i<-[p],
           el<-om2!!(i-1), tr<-trs ]++[norm(tr++[s*n,el],1)
           |i<-[p+1..lom2],el<-om2!!(i-1), tr<-trs ]
           ++[norm(qu++[n],0)|qu <- tuplesL 4 gon]
      else []
    s2=if q==1
      then[norm(tr++[el,s*n],1)| i<-[1..p],el<-om2!!(i-1),
           tr<-trs]++[norm(tr++[s*n,el],1)|i<-[p+1..lom2],
           el<-om2!!(i-1),tr<-trs]++[norm(qu++[n],0)
           | qu<-tuplesL 4 gon]
      else []
  in (newOrEmpty n chi (s1++s2))]
 where(firstRows,((gon,om2):lastRows))=splitAt(row-1)rows
       lom2 = length om2; trs = tuplesL 3 gon
```

For the function ext we assume $s = 1$ or $s = -1$, the sign of the new element n. For $q = 1$ we insert the new element at a new position p within the rank 2 oriented matroid. For $q = 0$ we insert the new element at an already existing position p within the rank 2 oriented matroid.

The function tailTup returns all new 5-tuples that occur the first time when we have n as the new element, that is, all 5-tuples with n at the end. At first the signs of all $\binom{n}{5}$ signed bases are considered to be unknown, that is, `signs` provided such a list with entries 2.

7.8 A 3-sphere with ten Dürer polyhedra

Sometimes it is just the beauty of mathematics that is a main motivation for studying certain geometrical objects. Platonic solids and many different generalizations of them have always attracted people because of their symmetries. At Dürer's time, mathematics and art were closely linked. We seldem dare to ask an artist where we find applications of his or her work. We can say that a mathematician is to some extent an artist as well. We have this connection in mind when we present the following sphere. We are going to describe its combinatorial facet types that we know from the famous copperplate engraving Melencolia I of Dürer, see Figure 7.2.

Figure 7.2   Dürer's copperplate engraving MELENCOLIA I from 1514

This masterpiece is probably the unique work of art in art history that has provoked most written material. In a profound study, in Schuster's two volume book from 1991, we have more than 30 pages of references of books and articles that are devoted just to this particular work of Dürer. However, the question that might interest a geometrician, *what are the metrical properties of this polyhedron?*, has been answered in a satisfactory manner only later, see Lynch, 1982. There are of course reconstructions from Dürer's perspective drawing, however, these findings coincide with Lynch's findings. We have depicted front view, top view, and side view of Dürer's polyhedron in Figure 7.3 according to T. Lynch. The relation to the magic square is apparent. Compare the radius of the circle in the top view with the circle of the ball and compare this with the edge length of the magic square. How is it that both midpoints together with the point where the rotational symmetry axis intersects the upper facet form three vertices of a regular pentagon? The reader might want to read more about Dürer's intentions. However, we have to come back to our 3-sphere that we have announced.

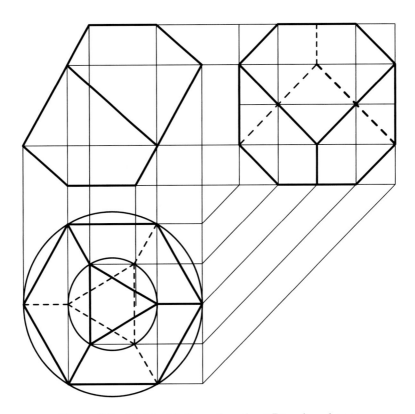

Figure 7.3   Metrical information about Dürer's polytope

Given simplicial 3-sphere that turned out to be a matroid polytope

5197	1978	1908	5139	1390	1340
1208	2089	2069	1240	2406	2456
2369	3690	3670	2356	3567	3517
3470	4706	4786	3417	4178	4128
4586	5867	5897	4528	5289	5239

Hyperline sequences of the matroid polytope

```
123|4 7-0 8 5-6 9    124|3 6 0-8-9-5-7   125|3 6-9 7 4-0 8   126|3 9-4 0-8-7-5   127|3 0-8 6-9-5 4   128|3 7 6 4-0-9-5
129|3 5 7 4-0 8-6    120|3 5 9 8-4-6-7   134|2 9 5 8 7-0-6   135|2 0 9-7-8-4-6   136|2 5 8-0 7 4-9   137|2 8 9 5-4-0-6
138|2 0-7 9 5-4-6    139|2 6 0-5-7-8-4   130|2 6 7 4-9-5-8   145|2 0-8 6 3-9 7   146|2-3-7 0-8-5-9   147|2 6 0 3-8-9-5
148|2 0 9 5 6 3 7    149|2 0-8 6 3 5 7   140|2-3-7-6-5-9-8   156|2-3 4 8-0 7-9   157|2 4 8 9-3-0-6   158|2 0-7 9-3-6-4
159|2 6 0 3-7-8-4    150|2 4 6 7-9 3-8   167|2 9 5-4 3 0-8   168|2 0 9 5-4 3 7   169|2-3-5-7 0-8-4   160|2-7-3 4-5-9-8
178|2 0 3 5 9-4-6    179|2 6 0 3 5-8-4   170|2 6-3 4-5-9-8   189|2 4 6 3 5 7-0   180|2 4 6 7 3 5 9   190|2 4 6 7 5 3-8
234|1-7 0-5-8 6-9    235|1 8 4 7 6-9-0   236|1 0 9-5-7-8-4   237|1 4 0-5 6-9-8   238|1-5 4 6-9 7-0   239|1 4 8 7 5-6-0
230|1 8 5-6 9-7-4    245|1 9 8-6-7-3-0   246|1 9 3 7 8 5-0   247|1 9-6 8 5-0-3   248|1-5-6-7-3-9-0   249|1-5-7-3-6 8-0
240|1 8 9 5 3 7 6    256|1 0 9 3-4-8-7   257|1 0 3-4-8 6-9   258|1 3 7 6 4-9-0   259|1 7 4 8-3-6-0   250|1 8-3-6 9-7-4
267|1-5 4 8-3-9-0    268|1-5 4-7-3-9-0   269|1 4 8 7 5 3-0   260|1 8 7 5 3 9-4   278|1-5 4 6-9-3-0   279|1-5 4 8-3-6-0
270|1 8-6 9 5 3-4    289|1 0-5-7-3-6-4   280|1-9-5-3-7-6-4   290|1 8-6-3-5-7-4   345|1 9-7 0 2-8 6   346|1-7 0-8-5-2-9
347|1 8 9 5 2 6 0    348|1-7 0 9 2 5 6   349|1-5-7 0-8 6 2   340|1-7-6-2-5-9-8   356|1 4 8 7-2-9-0   357|1-6-2-0-4-8-9
358|1-6-4-2-0 7-9    359|1 7 8 4 0 6 2   350|1 8 4 7-9 6 2   367|1 8 9 2 5-0-4   368|1 0-7 9 2 5-4   369|1 0-2-5-7-8-4
360|1 4 7-9-2-5-8    378|1-5-9 4 0 6 2   379|1-5 8 4 0 6 2   370|1 4-6-2-5-9-8   389|1-5-7 0 4 6 2   380|1-5-9 4 7 6 2
390|1 8 4 7 5 2 6    456|1 0 9 2-8-7-3   457|1 3 0 2-8 6-9   458|1 3 7 6-2-9-0   459|1 7-2 8-6-0-3   450|1 8-6 9-7-3 2
467|1 3 0-8-5-2-9    468|1 3 7-5-2-9-0   469|1-2 3 7 8 5-0   460|1 8 5 9 2-7-3   478|1-6-5-2-9-0-3   479|1-5-2-6 8-0-3
470|1 8 9 5 2 6-3    489|1 0-5-6-7-3 2   480|1-9-5-6-7-3 2   490|1 8-6-5-7-3 2   567|1 9 2 4 8-3-0   568|1 0 9 2 4-7-3
569|1 0-2 3-4-8-7    560|1 7-9-2 3-4-8   578|1 3 0 2 4 6-9   579|1 3 0 6 2 4 8   570|1-6-2 3-4-8-9   589|1 3 0 6 4 2-7
580|1 3 7-9 6 4 2    590|1 7 8 4-3 2 6   678|1 4-5-2-9-3-0   679|1-5 4 8-3 2-0   670|1 8 2 9 5 3-4   689|1 0-5 4-7-3 2
680|1-9-5 4-3-7 2    690|1 8 4 2-3-5-7   789|1 3 0 4 6 2 5   780|1-5-9 4-3 6 2   790|1-5 8 4-3 2 6   890|1-2-4-6-5-7-3
```

Its symmetry and the topological representation of its dual sphere can compete with many other geometrical objects. Perhaps it is just the fact that it cannot be represented via a polytope that makes it interesting. This 3-sphere has turned out to be a smallest non-polytopal matroid sphere with respect to its high symmetries, see Bokowski and Schuchert, 1995a.

There is of course an additional motivation for studying this sphere. We use it as a sample. We exemplify how an abstract sign structure can be interpreted topologically. This understanding helps us to obtain additional ideas when an open problem demands new creative ways of tackling it.

The 3-sphere that we are going to describe in more detail occurs in the Folkman–Lawrence representation of a simplicial matroid polytope with the facets that were given above.

We will later show that there is no corresponding 4-polytope that has the given simplicial sphere as its face lattice. Note that within the topological representation of the matroid polytope, the vertices become topological hyperplanes. The original simplicial sphere and the one from the Folkman–Lawrence representation form a dual pair. We obtain the second face lattice from the first, and vice versa, by putting the face lattice upside down.

We provide not only the list of hyperline sequences (rank 2 contractions) but also the complete list of circuits and the complete list of cocircuits that describe the matroid polytope.

All circuits of the matroid polytope

```
      1234567890      1234567890      1234567890      1234567890      1234567890

  1(+---++0000)   2(++---0+000)   3(+--++00-00)   4(++--+000-0)   5(+++--0000-)   6(+---0++000)   7(+---0+0+00)
  8(+++-0-00-0)   9(+---0+000-)  10(++--00+-00)  11(++--00+0-0)  12(+---00+00+)  13(+++-000+-0)  14(+++-000-0-)
 15(+++-0000--)  16(+--0++-000)  17(+--0++0-00)  18(+++0--00-0)  19(+++0--000-)  20(++-0+0+-00)  21(++-0+0+0-0)
 22(+++0-0-00-)  23(++-0+00--0)  24(++-0+00-0-)  25(+++0-000+-)  26(++-00-+-00)  27(+++00--0-0)  28(+-+00+-00-)
 29(+++00-0--0)  30(+++00-0-0-)  31(+---00++0-)  32(++-000++-0)  33(+++000--0-)  34(+++000-0--)  35(++-0000+--)
 36(+-0+++-000)  37(+-0++-0-00)  38(++0-+-00-0)  39(++0--+000-)  40(+-0+0++-00)  41(+-0++0-0-0)  42(++0--0+00-)
 43(++0-+00+-0)  44(++0-+00-0-)  45(++0-+000--)  46(+-0-0++-00)  47(++0-0-+0-0)  48(+-0-0+-00-)  49(++0-0+0+-0)
 50(++0-0+0-0-)  51(++0-0+00--)  52(++0-00++-0)  53(++0-00+-0-)  54(++0-00+0--)  55(++0-000+--)  56(+-00-++00-)
 57(++00+--0-0)  58(+-00++-00-)  59(++00+0-0--)  60(++00+-0--0)  61(+-00-++0-0)  62(+-00+0-+-0)  63(++00+0-0-0)
 64(++00+0-0--)  65(++000-0-+-)  66(++000-0+--)  67(++000+0-+-)  68(+-000-0+-)   69(++000-0--+)  70(++0000--+-)
 71(+0---++000)  72(+0---+0+00)  73(+0-+-+000-)  74(+0---+000-)  75(+0-++0+-0)   76(+0-++0-0-)   77(+0---+00+)
 78(+0-++00--0)  79(+0-++00-0-)  80(+0--+000-+)  81(+0--0++-00)  82(+0--0+-0-0)  83(+0--0-0++)   84(+0--0+0+-0)
 85(+0--0+0-0-)  86(+0--0+00--)  87(+0--+00++0)  88(+0--00-0++)  89(+0--00-0+-)  90(+0--000+-)   91(+0--0+++00)
 92(+0-0++-0-0)  93(+0+0-+-00-)  94(+0-0+0++--)  95(+0-0+0-0-)   96(+0+0-00++-)  97(+0-0+0+-0-)  98(+0-0+0+0+-)
 99(+0-0+0-0-+) 100(+0-0+00--+) 101(+0-00++-+0) 102(+0+00+--0-) 103(+0+00+0---) 104(+0-00+0+--) 105(+0-000-++)
106(+00-+-+00-) 107(+00++--0-0) 108(+00--+-00-) 109(+00++-0--0) 110(+00++-0-0-) 111(+00--+00-+) 112(+00-+0-+-0)
113(+00++0---0) 114(+00++0-0--) 115(+00+-00+-0) 116(+00-0+-+-0) 117(+00+0-+--0) 118(+00-0+-0--) 119(+00+0-0-0-)
120(+00+00--+-) 121(+000+--+-0) 122(+000++-0-)  123(+000++-0--) 124(+000+0-+--) 125(+000+0-+--) 126(+0000+--+-)
127(0-++++000)  128(0-+++-0-00) 129(0+++--00-0) 130(0++++-000-) 131(0-+++0--00) 132(0-+++0-0-0) 133(0++++-0-00)
134(0---++00-+0)135(0--++00-0+) 136(0----+000-) 137(0-+-0+-+00) 138(0---0++0+0) 139(0+--0-0++)  140(0---0+0++0)
141(0---0+0+0+) 142(0---+0+00+-)143(0---+00++0) 144(0---00++0+) 145(0---00+0++) 146(0++-000++-) 147(0+-0--+-+0)
148(0++0--+0-0)149(0++0--+00-) 150(0++0-0-+0-) 151(0++0-0+0-0) 152(0--0++00+-) 153(0-+0+0-+0-) 154(0++0-0-0+0)
155(0++0-0+0+-)156(0++0-0+0++) 157(0--0-0++-+0)158(0+00-0-+0+) 159(0+00-0-0-+) 160(0+00--0---+)161(0++000---+)
162(0+0+0--++00)163(0-0+++-0+0)164(0-0+++-00+) 165(0-0++-0-+0) 166(0-0++-0-0+) 167(0+0+0-+-00-)168(0-0++0-0+-)
169(0-0++00--0+)170(0-0++0-0-+)171(0-0+-00-+-) 172(0-0-0+-++0) 173(0-0-0++-+0) 174(0-0+0+-0+-) 175(0-0+0+0-+-)
176(0-0+00---+-)177(0+00--+--0)178(0+00--+-0-) 179(0+00+--0-+) 180(0+00+-0---+)181(0-00+0-+-+) 182(0-000+-+++)
183(00++--+-+00)184(00----++0+0)185(00--+-+00+)186(00---++0+0) 187(00---0+0+0) 188(00++--++0+-)189(00+-0-++0)
190(00---0++0+)191(00---0+0++) 192(00--++00++) 193(00--0+-++0) 194(00--0-++0)  195(00--0-0+++) 196(00--0+0++)
197(00--00++-+)198(00+0+---+0) 199(00+00-+-+0) 200(00+0--++0+-)201(00+0-0+++-) 202(00-0-0+-++) 203(00-0--+-++)
204(000++-+--0)205(000++-+-0-) 206(000++--0-+) 207(000++-0--+) 208(000-0+-+-+) 209(000-0++-++) 210(0000-++-+-)
```

We present all these data sets because it might be forgotten that behind our interpretations there are often rather long computations. In addition we show the different data structures for this example for comparing these data sets. Do these data sets contain beautiful aspects? It is the corresponding maximal cell in the Folkman–Lawrence representation that has a nice boundary structure.

All cocircuits of the matroid polytope

```
     1234567890        1234567890        1234567890        1234567890

   1(0000+-+++-)    2(000-0---++)    3(000++0++--)    4(000-+-0++-)    5(000-+--0++)    6(000--+--0+)    7(000-++--0)
   8(00+00++--+)    9(00-0-0---+)   10(00+0-+0--+)   11(00+0+++0++)   12(00+0+++-0+)   13(00-0-----0)   14(00-+00++--)
  15(00-+0-0++-)   16(00--0--0++)   17(00+-0+--0+)   18(00++0++--0)   19(00+-+00-++)   20(00+-+0+0++)   21(00---0--0+)
  22(00-+-0---0)   23(00+-+-00+)    24(00+-++0-0+)   25(00-+-+0--0)   26(00+++++00-)   27(00+++++0+0)   28(00+++++-00)
  29(0-000--++-)   30(0+00+0+++-)   31(0-00--0---)   32(0-00--+0--)   33(0-00+-++0-)   34(0+00+++++0)   35(0+0-00--++)
  36(0+0+0+0+++)   37(0+0+0+-0++)   38(0-0-0---0-)   39(0-0-0----+0)  40(0-0+-00--+)   41(0-0+-0+0--)   42(0+0++0++0-)
  43(0+0-+0-++0)   44(0-0-+-00+-)   45(0-0-+-0-0-)   46(0+0-++0+0+0)  47(0-0-+--+00-)  48(0-0-+--0+0)   49(0-0------00)
  50(0-+000++--)   51(0--00-0++-)   52(0+0+-00--0++)  53(0+0+00+--0+)  54(0-00+++---0)  55(0+0-0+00++-)  56(0+0+0-0+0++)
  57(0+-0-0--0+)   58(0-+0-0+--0)   59(0++0+00++0+)  60(0+++0+0-0+)   61(0-+0-0---0)   62(0-0+++++00-)  63(0-0+++0+0+0)
  64(0-0++++-00)   65(0-+-000--+)   66(0-0+-00+0--)  67(0+-+00+0+-)   68(0+-+00-++0)   69(0---0-00++-)  70(0---0-0-0-)
  71(0+-+0+0++0)   72(0---0-+00+)   73(0---0--0+0)   74(0-+-0---00)   75(0++-+000+0)   76(0-+-+00-0+)   77(0---+-00--0)
  78(0-+-+0+00-)   79(0-+-+0+0+0)   80(0-+-+0+-00)   81(0-----000-)   82(0-+-+-00+0)   83(0-+-+-0-00)   84(0+++++000)
  85(+0000--+++)   86(+000-0--++)   87(+000+-0++-)   88(+000---0+-)   89(+000---0++)   90(+000+--++0)   91(+00+00++++)
  92(+00+0-0+++)   93(+00-0--0++)   94(+00+0+++0+)   95(+00+0+++-0)   96(+00+-00++-)   97(+00+-0-0++)   98(+00++0++0+)
  99(+00+0++-0)   100(+00+-+00-+)  101(+00+-+0+0+)  102(+00++-0++0)  103(+00+-+--00+)  104(+00++0-0)   105(+00++-++00)
 106(+0+000++++)  107(+0+00-0+++)  108(+0+00++0++)  109(+0+00++-0+)  110(+0-00--++0)  111(+0+0-00-++)  112(+0+0-0+0++)
 113(+0-0-0--0+)  114(+0+0+0+++0)  115(+0+0-0--00+)  116(+0+0-+0-0+)  117(+0+0+-0+0+0)  118(+0-0---00+)  119(+0-0---0-0)
 120(+0-0---++00)  121(+0+-000--++)  122(+0+-00+0++)  123(+0-+00+0--)  124(+0+0+-0-00+)  125(+0+-0-00++)  126(+0+-0+0-0+)
 127(+0-+0-0+00)  128(+0++0++00+)  129(+0+0+0+0-0)  130(+0-+0-++00)  131(+0++-000+)  132(+0-+-00+0+)  133(+0-+-00+-0)
 134(+0-+--00+)  135(+0-+-0-0-0)  136(+0++++0+00)  137(+0++-+000+)  138(+0-+-+00-0)  139(+0-+--0+00)  140(+0++++++000)
 141(++0000--++)  142(+-000-0++-)  143(++000--0++)  144(+-000-++0-)  145(+-000---+0)  146(++00+00++--)  147(++00+0-0++)
 148(+-00-0--0+)  149(++00+0-++0)  150(+-00--00--)  151(+-00--0+0-)  152(++00+0+++0)  153(+-00---00+)  154(+-00---0-0)
 155(+-00---++0)  156(++0+000+++)  157(++0+0-0+++)  158(+-0+00+0++)  159(+-00+--++0)  160(+-0-0--0+0)  161(+-0-0+0-0+)
 162(+-0+0-0++0)  163(+-0-0-+00+)  164(+-0-0--0+0)  165(+-0+0-++00)  166(+-0+-000-+)  167(+-0+-00+0+)  168(++0+00++0+)
 169(+-0+-0-00+)  170(+-0+-0+0-0)  171(++0+0+++00)  172(+-0----000-)  173(+-0+--00-0)  174(+-0+---000+)  175(+-0------000)
 176(+++0000-++)  177(+++000++0+)  178(+-+000++0+)  179(+-+000++--0)  180(+++00-00++)  181(+++00+0-0+)  182(+-+00-0++0)
 183(+-+00++0+)  184(+-+00++0-0)  185(+-+00-++0)  186(+++0+000++)  187(+-+-0-00-+)  188(+++00++0+0)  189(+-+-0+00+)
 190(+-+-0-0+0-)  191(+-+-0+0++00)  192(+-+-0----000)  193(+-+-0--00-0)  194(+-+-0--0+00)  195(+-+-0++++000)  196(++++0000++)
 197(+-+-000-0+)  198(+++-+000++0)  199(+-+-00+00+)  200(+-+-00+0-0)  201(+++-+00++00)  202(++++0+000+)  203(+-+-0-00+0)
 204(+-+-0-0-00)  205(+-+-0-+000)  206(+-++-0000+)  207(+--+-000-0)  208(++-+-0+00)  209(+-+-+0+000)  210(+-+----0000)
```

7.8.1 The dual 3-sphere with its Heegard splitting into two tori

When we investigate the dual simple 3-sphere, we have 30 vertices.

Each vertex can be written with the labels of the former facets as a 4-tuple.

We start with the description of the dual 3-sphere of the simplicial 3-sphere of this section by starting with their facets. The facets are combinatorially equivalent to Dürer's polyhedron in his copperplate engraving Melencolia I from 1514, see Figure 7.2. For a first half of our sphere we take five copies of this polyhedron and we glue a first top triangle with a bottom triangle and so on. When we do this five times in a cyclic order, we obtain a torus which can be viewed as being subdivided into five combinatorial copies of Dürer's polyhedron. In addition, we identify the ith pentagon in the marked grey strip of pentagons in Figure 7.4 with the $(i+5)$th pentagon so that we again get a solid torus. We have gone around twice. We have depicted the topological structure of the cell decomposition in Figure 7.6. The red Möbius strip splits the depicted torus so that the original chain of five Dürer polyhedra becomes apparent.

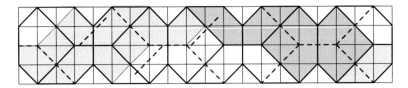

Figure 7.4   Five copies of Dürer's polyhedron glued along their triangles

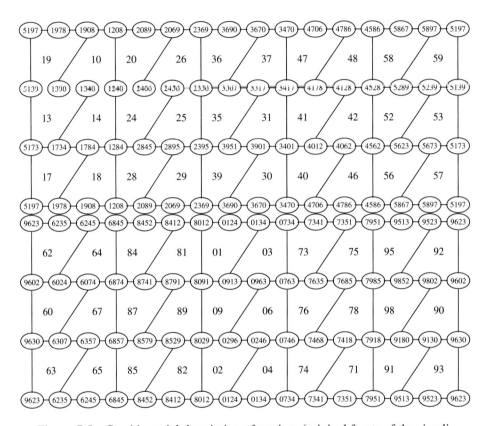

Figure 7.5   Combinatorial description of vertices (original facets of the simplicial 3-sphere) and subfacets (some original edges) defining the Heegard-splitting

A model can often show more than any sophisticated computer graphic. Figure 7.7 shows such a pottery model and a fake way of gluing it to a second. Here the second torus is just a photocopy of the first. A first correct way of gluing these two tori can be seen in Figure 7.8.

The second version of such a solid subdivision of the 3-torus into five copies of Dürer's polyhedron has the same combinatorial boundary structure. If we

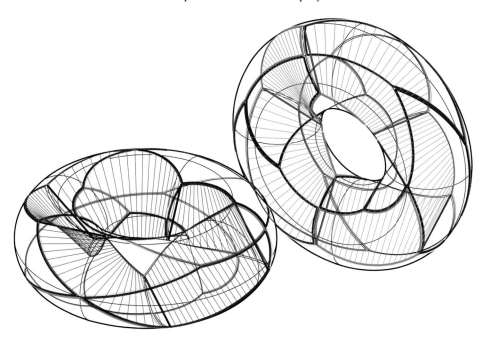

Figure 7.6   Two solid tori with five Dürer polyhedra, a Möbius band consisting of five pentagons, can be seen in the interior of the torus; computer graphics by E. Hartmann, Darmstadt

glue these two solid tori along their common boundary, we obtain a 3-sphere. This boundary then has become a Heegard splitting of the 3-sphere, compare Page 140.

We try to understand the symmetry better. We look at Figure 7.9.

Here we see the 2-torus of the Heegard splitting within a square. We label the yellow patches from light to dark as 1,2,3,4,5 and the blue patches from light to dark as 6,7,8,9,0. We obtain a periodical double covering of the plane, a yellow covering and a blue one. Each vertex of the tiling is incident to precisely four patches. The fundamental region, marked as a black square, defines a torus if we identify opposite edges of the square. This torus has been depicted with different colours for the pentagons in Figure 7.10. We see additional white and black edges. These additional white edges lie in the interior of the first solid torus, additional black edges lie in the interior of the second solid torus. Observe that this leads to five Dürer polytopes in each case when we add a further five white edges and five black edges in the interior that are each covered by two other edges. Try first to imagine the white edges as showing five combinatorial Schlegel diagrams of

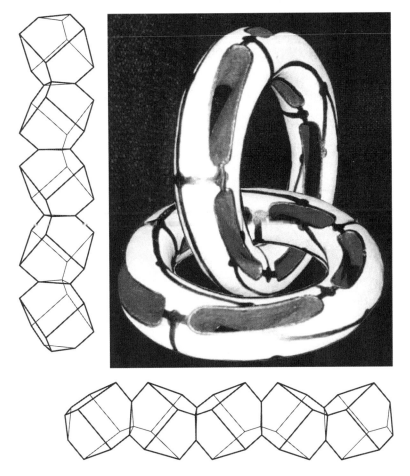

Figure 7.7   Two sequences of five Dürer polyhedra, each glued together to form a torus depicted here as a pottery model

cubes each having cut off two opposite vertices. Then you see the same structure again when you use the black edges.

A translation (from light to dark) and a rotation of the black square maps the 3-sphere onto itself. We obtain symmetries expressed via permutations $a = (1, 2, 3, 4, 5)(6, 7, 8, 9, 0)$ and $b = (1, 0, 2, 8)(9, 4)(6, 5, 4, 3)$.

We try to find all uniform oriented matroids that are compatible with the given simplicial 3-sphere and that have the same symmetry group. Assuming that such an oriented matroid exists, we have orbits of signed bases and orbits of hyperline sequences under the symmetry group. We have eight orbits of hyperline sequences and sixteen orbits of signed bases.

The oriented matroid of rank 5 that we have found for this equi-facetted 3-sphere can be described in terms of its Folkman–Lawrence representation. We

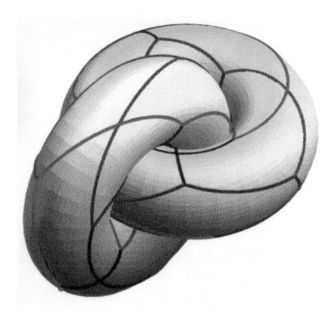

Figure 7.8 Two solid tori with ten cells isomorphic to Dürer's polyhedron used in his Melencolia I; computer graphics by Susanne Hayn, Darmstadt

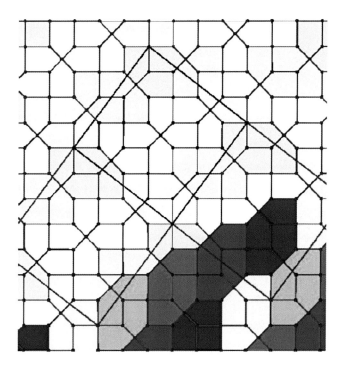

Figure 7.9 Description of an interesting 3-sphere

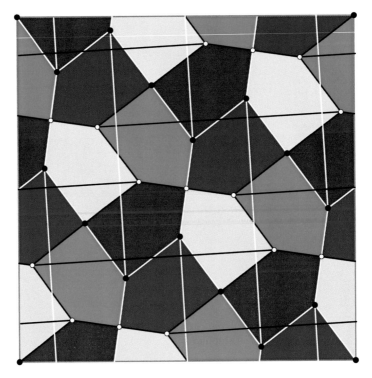

Figure 7.10 Fundamental region of the tiling of the plane in Figure 7.9. Description of all ten Dürer polyhedra

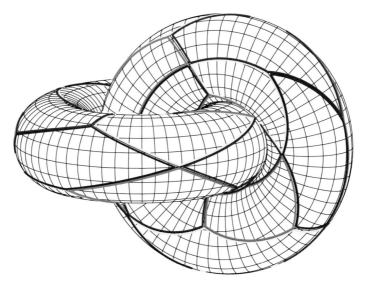

Figure 7.11 Both solid tori containing five Dürer polyhedra each; computer graphics by E. Hartmann, Darmstadt

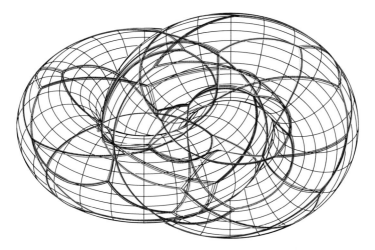

Figure 7.12 How the second torus touches the first one, computer graphics by E. Hartmann

interpret the chirotope data structure of our oriented matroid via the topological representation theorem of Chapter 4. We have ten topological hyperplanes in projective 4-space that bound the decisive maximal cell, the matroid polytope. We can assume that we have as a model for our 4-dimensional projective space a unit 4-ball with identified anti-podal points. The ten topological hyperplanes are three-dimensional topological balls that intersect this 4-ball. Antipodal points on the boundary of this 4-ball have to be identified. Each of these three-dimensional balls is a rank 3 contraction of the full oriented matroid. They each correspond to one facet of the matroid polytope. We can look at all these three-dimensional topological balls corresponding to the ten facets separately. They in turn have rank 3 contractions that correspond to subfacets of the original matroid polytope. In Figure 7.13 we see all these rank 3 contractions. The subfacets have been marked. Each facet has a color that supports the understanding. Above and on the right of a facet, we have all corresponding rank 3 contractions that we have to imagine within a topological 3-ball like in the model in Figure 1.24. Finally, all these ten topological 3-balls intersect pairwise nicely so as to form the boundary of the matroid polytope with ten Dürer polytope facets.

The matroid polytope cannot be realized as a convex polytope. The final argument hides the difficult process of finding this contradiction. It has turned out that the same final polynomial works for all compatible oriented matroids in this case, not for only the unique most symmetric one.

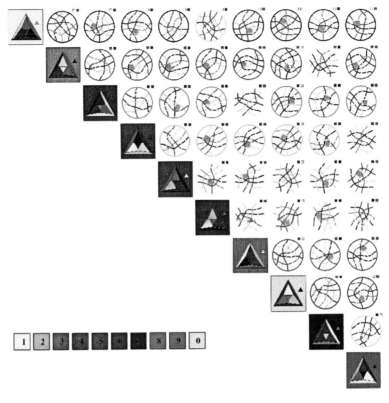

Figure 7.13   All rank 3 contractions of a rank 5 matroid polytope

We find the proof that this sphere cannot be realized as a 4-polytope in Bokowski and Schuchert, 1995. The method of proof will be shown in the next example.

7.9   Altshuler's sphere No. 425

3-sphere facets

```
sphereA425::[[Int]]
sphereA425
= [[1,0,4,5],[3,2,6,7],[5,4,8,9],[7,6,0,1],[9,8,2,3],
   [0,1,3,2],[2,3,5,4],[4,5,7,6],[6,7,9,8],[8,9,1,0],
   [1,0,4,8],[1,0,6,2],[0,1,3,5],[0,1,9,7],[3,2,6,0],
   [3,2,8,4],[2,3,5,7],[2,3,1,9],[5,4,8,2],[5,4,0,6],
   [4,5,7,9],[4,5,3,1],[7,6,0,4],[7,6,2,8],[6,7,9,1],
   [6,7,5,3],[9,8,2,6],[9,8,4,0],[8,9,1,3],[8,9,7,5],
   [8,2,7,5],[0,4,9,7],[2,6,1,9],[4,8,3,1],[6,0,5,3]  ]
```

There is another simplicial matroid polytope with ten elements and 35 facets that leads to an interesting Folkman–Lawrence representation. The sphere has appeared in Altshuler's list of 3-spheres as No. 425. The polar dual 3-sphere that shows up in this representation has again only one combinatorial facet type.

We have depicted the facets of this polar dual sphere on the main diagonal in Figure 7.14. As in Figure 7.13, we see all rank 3 contractions of the rank 5 matroid polytope. The subfacets appear as 2-cells in the rank 3 contractions and they have been marked with a red dot. This representation of Figure 7.14 has been published first in Bokowski, 1994.

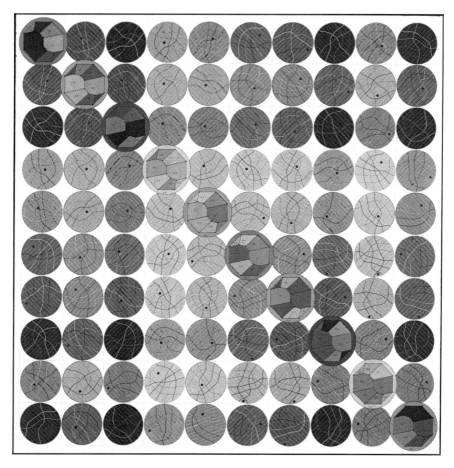

Figure 7.14 Folkman–Lawrence representation of Altshuler's 3-sphere M_{425}^{10}. Schlegel diagrams of facets on the main diagonal. All rank 3 contractions with a dot in a cell marking a subfacet

7.9.1   *Non-polytopality proof*

Equal simplex orientations according to supporting hyperplanes

```
4  4  4  4  4  4  4-8-2  2  2  2  2  2  2  2  2  2  2  2  2  2  2-6-0  0  0  0
1  1  1-7  7  7  7  7  7-3  3  3  3  3  3-9  9  9  9  9  9-5  5  5  5  5  5-1  1  1
0  0  0  0  0  0  0  0  0  0  0  0  0  0-4-8  8  8  8  8  8  8  8  8  8  8  8  8  8  8
8-2-6  6  6  6  6  6  6  6  6  6  6  6  6  6  6  6  6  6-0-4  4  4  4  4  4  4  4  4
5  5  5  5-2-8-1  1  1  1-4-8-7  7  7  7-4-0-3  3  3  3-0-6-9  9  9  9-2-6
```

The reason why this sphere cannot be polytopal can be argued as follows. The orientations of 4-simplices that are written as columns in the above scheme have to be equal. The elements that are not connected horizontally have to lie on the same side of the corresponding facet that remains when these elements are discarded.

Each Grassmann–Plücker relation between determinants of a matrix that contains the homogeneous coordinates of a corresponding polytope of the sphere is zero. The same holds for the following linear combination of such.

Linear combination of Grassmann–Plücker relations

```
+{680|1742}[26034][24085][26894]
+{260|1348}[68074][24085][26984]
+{246|3708}[26809][24085][68041]
+{268|9704}[26034][24085][68041]
+{280|9346}[24856][46072][68041]
-{248|3506}[26809][46072][68041]
+{468|9520}[48012][46072][26038]
+{480|9162}[24856][46072][26038]
+{240|1568}[68074][26894][26038]
-{460|5782}[48012][26894][26038]  =  0
```

After evaluating the Grassmann–Plücker relations according to the formula from Page 44 and after deleting corresponding monomials, we obtain the following identity between determinants.

Final polynomial that shows non-polytopality

```
-[02468][68017][26034][24085][26894]
+[02468][26013][68074][24085][26984]
-[02468][24637][26809][24085][68041]
-[02468][26897][26034][24085][68041]
+[02468][28093][24856][46072][68041]
-[02468][24835][26809][46072][68041]
-[02468][46895][48012][46072][26038]
+[02468][48091][24856][46072][26038]
+[02468][24015][68074][26894][26038]
+[02468][46057][48012][26894][26038] = 0
```

When we now plug in the signs that we know from the above observation, we find ten monomials the signs of which are all equal. This is a contradiction to the assumption that we can find coordinates. The 3-sphere is not polytopal, compare Bokowski and Garms, 1987.

The sphere has the group $\mathbb{Z}_5 \times \mathbb{Z}_4$ as its combinatorial symmetry group. This symmetry can be used to reduce the argument for showing non-polytopality. We have not worked this out. The reader can find this symmetry argument in Bokowski and Richter-Gebert, 1990.

When this polynomial is known, you can easily argue that the sign structure is not consistent with the polynomial. We have a similar argument for proving the projective incidence theorem of Pappus on Page 267. Finding such a polynomial is in general a long process.

A more detailed investigation can be found on Page 213.

7.10 A 3-sphere of Gévay, self-polar-dual example

The extension algorithm for finding matroid polytopes in the non-uniform case can be applied in case of a self-polar-dual sphere that occurred independently several times. A series of self-polar-dual 3-spheres that includes the 24-cell was found by G. Gévay. We deal here with the smallest example that has 15 vertices. The same example was found by F. Santos and G. Ziegler, independently. In Paffenholz, 2004, we find coordinates for a corresponding polytope and more aspects and references about this class of polytopes. We have an example where a direct method was sucessful for finding coordinates.

The sphere can be described by the list of facets and subfacets as indicated by the following function `ex`. There is a symmetry group of order 6 with the following generators.

$$x = (1, 2, 3)(4, 5, 6)(7, 8, 9)(10, 11, 12)(13, 14, 15)$$

$$y = (1, 4)(2, 5)(3, 6)(7)(8)(9)(10, 14)(12, 13)(11, 15)$$

Finding a matroid polytope such that its face lattice coincides with the given one was used as an example for writing the Haskell functions for Section 7.3 and that which has followed.

3-sphere of Gévay, Santos, Ziegler, Paffenholz

```
ex::[(Int,[[Int]])]   -- facet number with oriented triangles
ex=[( 1,[[1,10,7],[1,13,10],[1,7,13],[3,7,10],[3,10,13],[3,13,7]]),
    ( 2,[[2,11,8],[2,14,11],[2,8,14],[1,8,11],[1,11,14],[1,14,8]]),
    ( 3,[[3,12,9],[3,15,12],[3,9,15],[2,9,12],[2,12,15],[2,15,9]]),
    ( 4,[[4,14,7],[4,12,14],[4,7,12],[6,7,14],[6,14,12],[6,12,7]]),
    ( 5,[[5,15,8],[5,10,15],[5,8,10],[4,8,15],[4,15,10],[4,10,8]]),
    ( 6,[[6,13,9],[6,11,13],[6,9,11],[5,9,13],[5,13,11],[5,11,9]]),
    ( 7,[[1, 7,10],[1,10, 8],[1, 8,14],[1,14, 7],
         [4,10, 7],[4, 8,10],[4,14, 8],[4, 7,14]]),
    ( 8,[[2, 8,11],[2,11, 9],[2, 9,15],[2,15, 8],
         [5,11, 8],[5, 9,11],[5,15, 9],[5, 8,15]]),
    ( 9,[[3, 9,12],[3,12, 7],[3, 7,13],[3,13, 9],
         [6,12, 9],[6, 7,12],[6,13, 7],[6, 9,13]]),
    (10,[[1,13, 7],[1, 7,14],[1,14,11],[1,11,13],
         [6, 7,13],[6,14, 7],[6,11,14],[6,13,11]]),
    (11,[[2,14, 8],[2, 8,15],[2,15,12],[2,12,14],
         [4, 8,14],[4,15, 8],[4,12,15],[4,14,12]]),
    (12,[[3,15, 9],[3, 9,13],[3,13,10],[3,10,15],
         [5, 9,15],[5,13, 9],[5,10,13],[5,15,10]]),
    (13,[[4,12, 7],[4, 7,10],[4,10,15],[4,15,12],
         [3, 7,12],[3,10, 7],[3,15,10],[3,12,15]]),
    (14,[[5,10, 8],[5, 8,11],[5,11,13],[5,13,10],
         [1, 8,10],[1,11, 8],[1,13,11],[1,10,13]]),
    (15,[[6,11, 9],[6, 9,12],[6,12,14],[6,14,11],
         [2, 9,11],[2,12, 9],[2,14,12],[2,11,14]]))]
```

The number of elements makes this problem more difficult. We have to deal with 3003 signs. The sphere is far from being rigid. When this chapter was written, the corresponding problem was handed over to D. Bremner who wrote a corresponding C-program. The input file of his program contains the point lists of all facets.

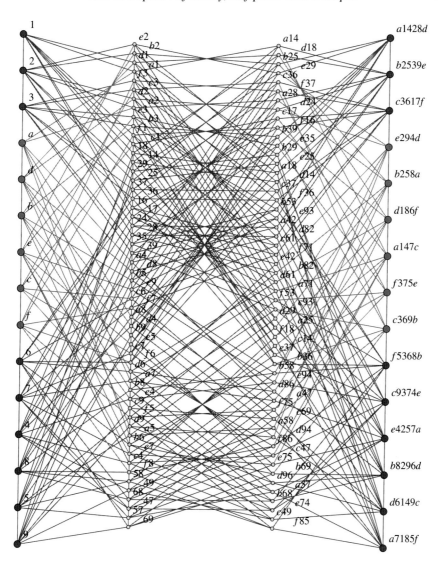

Figure 7.15 Face lattice of a self-polar-dual 3-sphere of Gévay, Paffenholz, Santos, and Ziegler

```
rank=5 elements=15
boundary={
          [1  7 10 13  3]  [4  7 14 12  6]  [2  8 11 14  1]
          [5  8 15 10  4]  [3  9 12 15  2]  [6  9 13 11  5]
     [1  4  7   8 10 14]  [2  5  8   9 11 15]  [3  6  7   9 12 13]
     [1  6  7 11 13 14]  [2  4  8 12 14 15]  [3  5  9 10 13 15]
     [1  5  8 10 11 13]  [2  6  9 11 12 14]  [3  4  7 10 12 15] }
```

As a result of this program, we have found many admissible chirotopes for this 3-sphere. In other words, there exist a lot of matroid polytopes with the given 3-sphere structure. So, it was no surprise that Paffenholz found coordinates during our independent investigation.

The face lattice structure in Figure 7.15 shows the self-duality property. A result that we also have for a 3-sphere of P. McMullen and that we consider in the next section. The existence of a corresponding polytope is unknown and the method of Paffenholz seems not to work in this case.

7.11   On a 3-sphere of McMullen

The higher dimensionsal analogues of the five Platonic solids are known from Schläfli's work in 1850, first published in 1900. We have in all dimensions the d-cube, the d-crosspolytope, and the d-simplex. In dimension 4 we have in addition the self-polar-dual 24-cell, the 120-cell, and the 600-cell. For historical remarks in this context up to abstract regular polytopes compare McMullen, 1994.

For the study of a problem of McMullen, it is useful to have a profound understanding of this last polar dual pair of regular polytopes, the 120-cell, and the 600-cell. The problem can be interpreted as lying between these two polytopes with a Coxeter symmetry group of order 14 400.

We start by describing the 600-cell. For the coordinates of the 600-cell, we use the number t defined as $t = 1/2 + 1/2\sqrt{5}$ which is a zero of the polynomial $t^2 - t = 1$, the relation that defines the golden section. The 600-cell is obtained as the convex hull of 120 points. In our case all points lie on a sphere of radius 2.

We prefer to provide the four-dimensional coordinates of all vertices in a geometrical manner. We use two orthogonal planar projections of all points, compare Figures 7.16 and 7.17 with identified centres. Taking into account both drawings, each point is obtained as the vector sum of its two projections.

With these projections some geometrical properties can be immediately understood. Each of the edges of the two large regular pentagons in both projections contains 12 projected points. They can be seen as projected icosahedra. The regular pentagon shape can be seen in the complementary projection by looking at corresponding labels. The convex hull of these 120 points forms the regular 600-cell.

When the reader wants to see other projections, he or she can use any dynamical drawing software such as Cinderella, see Richter-Gebert and Kortenkamp, 1999.

Table 7.1   *Coordinates of the 600-cell*

Pt.	coordinates				Pt.	coordinates				Pt.	coordinates			
00	0	t-1	t	1	40	1	0	t	t-1	80	t	t-1	-1	0
01	0	0	2	0	41	t-1	-1	t	0	81	t	0	1-t	1
02	0	1-t	t	-1	42	0	-t	1	1-t	82	1	1-t	0	t
03	0	-1x	t-1	-t	43	1-t	-t	0	-1	83	0	-1	t-1	t
04	0	-1	1-t	-t	44	-1	-1	-1	-1	84	-1	-1	1	1
05	0	1-t	-t	-1	45	-1	0	-t	1-t	85	-t	1-t	1	0
06	0	0	-2	0	46	1-t	1	-t	0	86	-t	0	t-1	-1
07	0	t-1	-t	1	47	0	t	-1	t-1	87	-1	t-1	0	-t
08	0	1	1-t	t	48	t-1	t	0	1	88	0	1	1-t	-t
09	0	1	t-1	t	49	1	1	1	1	89	1	1	-1	-1
10	-1	0	t	t-1	50	0	1-t	t	1	90	t	1	0	t-1
11	-1	0	t	1-t	51	1-t	-1	t	0	91	t	0	t-1	1
12	1-t	0	1	-t	52	-1	-1	1	-1	92	1	-1	1	1
13	0	0	0	-2	53	-1	1-t	0	-t	93	0	-t	1	t-1
14	t-1	0	-1	-t	54	1-t	0	-1	-t	94	-1	-t	t-1	0
15	1	0	-t	1-t	55	0	t-1	-t	-1	95	-t	-1	0	1-t
16	1	0	-t	t-1	56	t-1	1	-t	0	96	-t	0	1-t	-1
17	t-1	0	-1	t	57	1	1	-1	1	97	-1	1	-1	-1
18	0	0	0	2	58	1	t-1	0	t	98	0	t	-1	1-t
19	1-t	0	1	t	59	t-1	0	1	t	99	1	t	1-t	0
20	1-t	1	t	0	60	t	0	t-1	-1	a0	t	t-1	1	0
21	0	t-1	t	-1	61	t	-1	0	1-t	a1	t	1-t	1	0
22	t-1	0	1	-t	62	1	-t	1-t	0	a2	1	-t	t-1	0
23	1	1-t	0	-t	63	0	-t	-1	t-1	a3	0	-2	0	0
24	1	-1	-1	-1	64	-1	-1	-1	1	a4	-1	-t	1-t	0
25	t-1	-1	-t	0	65	-t	0	1-t	1	a5	-t	1-t	-1	0
26	0	1-t	-t	1	66	-t	1	0	t-1	a6	-t	t-1	-1	0
27	1-t	0	-1	t	67	-1	t	t-1	0	a7	-1	t	1-t	0
28	-1	t-1	0	t	68	0	t	1	1-t	a8	0	2	0	0
29	-1	1	1	1	69	1	1	1	-1	a9	1	t	t-1	0
30	t-1	1	t	0	70	t	0	1-t	-1	b0	t	1	0	1-t
31	1	0	t	1-t	71	t	1-t	-1	0	b1	2	0	0	0
32	1	-1	1	-1	72	1	-1	-1	1	b2	t	-1	0	t-1
33	t-1	-t	0	-1	73	0	-1	1-t	t	b3	t-1	-t	0	1
34	0	-t	-1	1-t	74	-1	1-t	0	t	b4	1-t	-t	0	1
35	1-t	-1	-t	0	75	-t	0	t-1	1	b5	-t	-1	0	t-1
36	-1	0	-t	t-1	76	-t	t-1	1	0	b6	-2	0	0	0
37	-1	1	-1	1	77	-1	1	1	-1	b7	-t	1	0	1-t
38	1-t	t	0	1	78	0	1	t-1	-t	b8	1-t	t	0	-1
39	0	t	1	t-1	79	1	t-1	0	-t	b9	t-1	t	0	-1

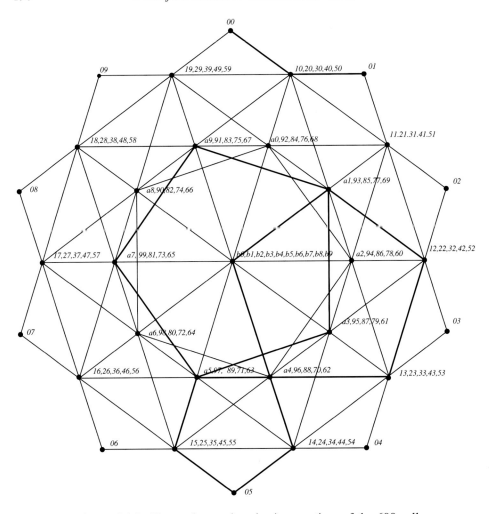

Figure 7.16   First orthogonal projection, vertices of the 600-cell

Pick two orthogonal unit vectors in each of the two orthogonal projections. Draw both projections with identified midpoints as in Figure 7.18 such that the projected unit vectors can be changed later with respect to length and direction. The original circles become ellipses.

Observe that any parallel projection of the 600-cell onto a plane shows the projected four unit vectors somewhere. When we drag the four unit vectors in an arbitrary position in the plane, we have defined the image under a parallel projection onto the plane. The former two projections are defined as well and we obtain all points in the projection as corresponding vector sums.

In Figures 7.19, 7.20, and 7.21 we have depicted three examples of such projections. We have marked the 12 circles that contain ten vertices each. Because

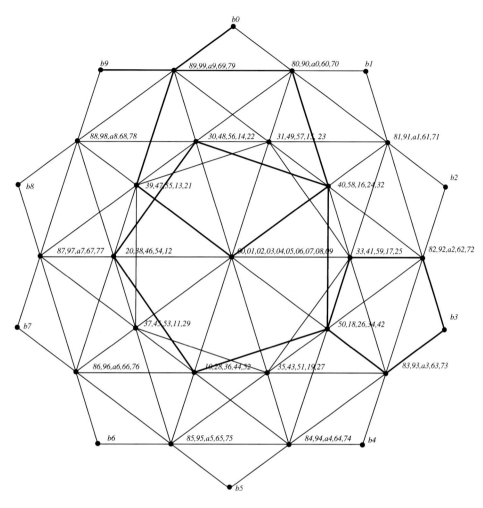

Figure 7.17   Second orthogonal projection, vertices of the 600-cell

of the symmetry of the 600-cell, we have altogether 12 projections of the two projection types with which we started.

We suggest that the reader play with the software Cinderella to get an additional insight of a geometrical flavor.

7.11.1   The polyhedral 240-cell and McMullen's self-polar-dual sphere

The coordinates of the 600-cell can be used to describe another non-convex star-shaped four-dimensional polyhedron that we call 240-cell. Take one of the above twelve circles C_i. Project all 120 vertices onto a plane parallel to the affine hull of this circle C_i along the two-dimensional orthogonal space V_i. A projection is

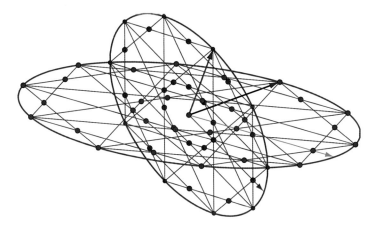

Figure 7.18   Planar affine images of projections of all vertices of the 600-cell

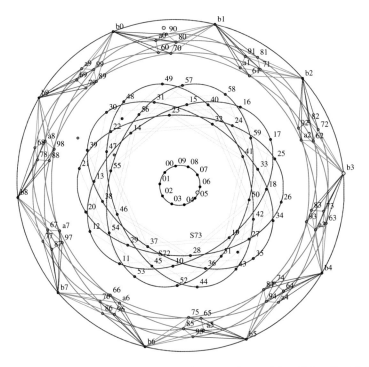

Figure 7.19   A first planar projection of the vertices of the 600-cell with Hopf fiber circles of the 3-sphere

obtained that is up to the labels equal to those of Figures 7.16 and 7.17. Now take the minimal closed star shaped polygon P_i in this projection that we have depicted in Figures 7.16 and 7.17 and that contains all projected vertices. It defines a four-dimensional cylinder $P_i + V_i$ with a two-dimensional axis when you take the Minkowski sum of P_i with the space V_i. The intersections of all of these

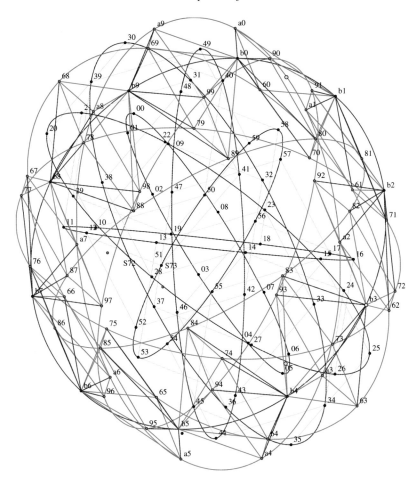

Figure 7.20 A second planar projection of the vertices of the 600-cell with Hopf fiber circles of the 3-sphere

twelve star-shaped minimal cylinders contain all 120 vertices. The 240 pyramids lie on the boundary of this intersection. They define a polyhedral 3-sphere, the 240-cell. This polyhedron occurred during Bokowski's and Mock's investigation of McMullen's sphere that can be defined topologically by gluing together pairs of regular pentagons of the facets of the 240-cell. See McMullen, 1968 for his topological sphere and see Bokowski, Cara, and Mock, 1999 for both McMullen's sphere and the polyhedral 240-cell.

We can take the polar dual of these two 3-spheres to obtain altogether three interesting 3-spheres. McMullen's sphere is a self-polar-dual one and it is the most interesting sphere. We have mentioned these spheres because we do not know whether there exist corresponding convex polytopes with the same face lattice.

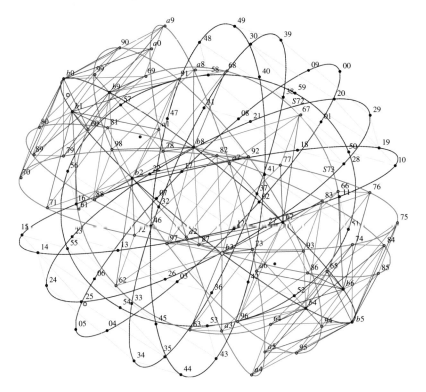

Figure 7.21   A third planar projection of the vertices of the 600-cell with Hopf fiber circles of the 3-sphere

We describe McMullen's sphere once more directly. For each edge of the 600-cell that occurs along the 12 regular 10-gons that we have marked in Figures 7.19, 7.20, and 7.21 as circles, McMullen replaces the five (tetrahedral) facets of the 600-cell that are incident with this edge with a combinatorial double pyramid over the pentagon. The former edge has to be deleted. It turns out that the combinatorial double pyramid over the pentagon remains as the only facet type of the constructed 3-sphere. We list all facets that form the 3-sphere in Table 7.2.

We finally show two rank 3 contractions of the 600-cell in Figure 7.22. A matroid polytope for McMullen's sphere has not been found so far. Is there a way to change topologically the supporting hyperplanes of the 600-cell to those of McMullen's sphere? Can we find a hyperplane configuration with an incidence structure that we would have in the case of a polytope? Both questions, if answered in the affirmative, do not answer completely McMullen's question. However, even those have not been answered so far, thus indicating that the question is a difficult one. We do have information about all subgroups of the face lattice, see Bokowski, Cara, and Mock, 1999:

Table 7.2 *Facets of McMullen's sphere with 120 double pyramids over pentagons and facets of the polyhedral 240-cell*

	pentagon			pentagon			pentagon	
00	10 20 30 40 50	01	40	01 31 a1 92 50	41	80	16 57 90 b1 71	81
01	11 21 31 41 51	02	41	02 32 a2 93 51	42	81	17 58 91 b2 72	82
02	12 22 32 42 52	03	42	03 33 a3 94 52	43	82	18 59 92 b3 73	83
03	13 23 33 43 53	04	43	04 34 a4 95 53	44	83	19 50 93 b4 74	84
04	14 24 34 44 54	05	44	05 35 a5 96 54	45	84	10 51 94 b5 75	85
05	15 25 35 45 55	06	45	06 36 a6 97 55	46	85	11 52 95 b6 76	86
06	16 26 36 46 56	07	46	07 37 a7 98 56	47	86	12 53 96 b7 77	87
07	17 27 37 47 57	08	47	08 38 a8 99 57	48	87	13 54 97 b8 78	88
08	18 28 38 48 58	09	48	09 39 a9 90 58	49	88	14 55 98 b9 79	89
09	19 29 39 49 59	00	49	00 30 a0 91 59	40	89	15 56 99 b0 70	80
10	01 51 85 76 20	11	50	10 01 41 93 84	51	90	58 49 a0 b1 81	91
11	02 52 86 77 21	12	51	11 02 42 94 85	52	91	59 40 a1 b2 82	92
12	03 53 87 78 22	13	52	12 03 43 95 86	53	92	50 41 a2 b3 83	93
13	04 54 88 79 23	14	53	13 04 44 96 87	54	93	51 42 a3 b4 84	94
14	05 55 89 70 24	15	54	14 05 45 97 88	55	94	52 43 a4 b5 85	95
15	06 56 80 71 25	16	55	15 06 46 98 89	56	95	53 44 a5 b6 86	96
16	07 57 81 72 26	17	56	16 07 47 99 80	57	96	54 45 a6 b7 87	97
17	08 58 82 73 27	18	57	17 08 48 90 81	58	97	55 46 a7 b8 88	98
18	09 59 83 74 28	19	58	18 09 49 91 82	59	98	56 47 a8 b9 89	99
19	00 50 84 75 29	10	59	19 00 40 92 83	50	99	57 48 a9 b0 80	90
20	01 11 77 68 30	21	60	23 70 b1 a1 32	61	a0	31 60 b1 91 40	a1
21	02 12 78 69 31	22	61	24 71 b2 a2 33	62	a1	32 61 b2 92 41	a2
22	03 13 79 60 32	23	62	25 72 b3 a3 34	63	a2	33 62 b3 93 42	a3
23	04 14 70 61 33	24	63	26 73 b4 a4 35	64	a3	34 63 b4 94 43	a4
24	05 15 71 62 34	25	64	27 74 b5 a5 36	65	a4	35 64 b5 95 44	a5
25	06 16 72 63 35	26	65	38 75 b6 a6 37	66	a5	36 65 b6 96 45	a6
26	07 17 73 64 36	27	66	29 76 b7 a7 38	67	a6	37 66 b7 97 46	a7
27	08 18 74 65 37	28	67	20 77 b8 a8 39	68	a7	38 67 b8 98 47	a8
28	09 19 75 66 38	29	68	21 78 b9 a9 30	69	a8	39 68 b9 99 48	a9
29	00 10 76 67 39	20	69	22 79 b0 a0 31	60	a9	30 69 b0 90 49	a0
30	01 21 69 a0 40	31	70	15 80 b1 61 24	71	b0	80 90 a0 60 70	b1
31	02 22 60 a1 41	32	71	16 81 b2 62 25	72	b1	81 91 a1 61 71	b2
32	03 23 61 a2 42	33	72	17 82 b3 63 26	73	b2	82 92 a2 62 72	b3
33	04 24 62 a3 43	34	73	18 83 b4 64 27	74	b3	83 93 a3 63 73	b4
34	05 25 63 a4 44	35	74	19 84 b5 65 28	75	b4	84 94 a4 64 74	b5
35	06 26 64 a5 45	36	75	10 85 b6 66 29	76	b5	85 95 a5 65 75	b6
36	07 27 65 a6 46	37	76	11 86 b7 67 20	77	b6	86 96 a6 66 76	b7
37	08 28 66 a7 47	38	77	12 87 b8 68 21	78	b7	87 97 a7 67 77	b8
38	09 29 67 a8 48	39	78	13 88 b9 69 22	79	b8	88 98 a8 68 78	b9
39	00 20 68 a9 49	30	79	14 89 b0 60 23	70	b9	89 99 a9 69 79	b0

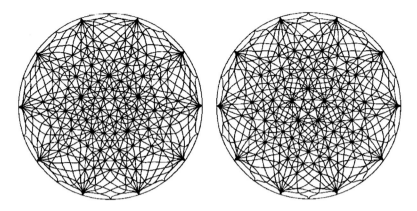

Figure 7.22   Two different rank 3 contractions of the 600-cell

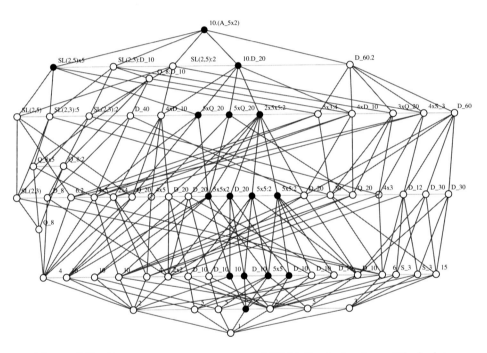

Figure 7.23   Symmetry group structure of McMullen's 3-sphere. Black dots mark those subgroups that cannot occur geometrically

Theorem 7.1 *[Bokowski, Cara, and Mock] All subgoups marked by a black dot in Figure 7.23 cannot occur as an affine symmetry of a polytope with a face lattice of McMullen's sphere, provided such a polytope does exist.*

This theorem is a stronger version of what was already known to McMullen when he wrote his Ph.D. thesis and which stimulated research in this direction:

if such a polytope existed, combinatorial and affine symmetries of its face lattice would differ, compare with Page 143.

7.12 Simple spheres with small number of facets

For a small number (≤ 10) of facets, all equifacetted simple 3-spheres were classified into polytopal and non-polytopal ones in Bokowski and Schuchert, 1995a, compare also Schuchert, 1995.

8

From oriented matroids to matrices

We have seen in the preceeding two chapters that arriving at an oriented matroid is a decisive first step for finding geometrical realizations of polytopes or other polyhedral objects. In this chapter we deal with the problem of finding a matrix for a given oriented matroid.

Problem 8.1 *Find for a given oriented matroid a corresponding matrix, representing it or a proof that such a matrix is impossible to find.*

On the one hand, there is a general algorithm to solve this problem. On the other hand, it is known that this algorithm from real algebraic geometry is far from applicable for our cases in the theory of oriented matroids. Consequently, we will deal with heuristic methods for small number of elements and small rank. A general algorithm for finding a realization for an oriented matroid is in principle available due to Collins's cylindrical algebraic decomposition method, see Collins, 1975 and see Basu, Pollack, and Roy, 2003. These methods from real algebraic geometry go far beyond the scope of this book. However, they are not applicable for the problem sizes in which we are interested.

For non-realizability proofs, see Bokowski, Richter-Gebert, and Sturmfels, 1990. For combinatorial criteria, see Richter-Gebert, 1989. A survey article about the topic of this chapter can be found in Bokowski, 2001. We are interested in fast heuristic algorithms to decide on the realizability of oriented matroids.

8.1 An example of a 5-sphere due to Shemer

The following problem turns out to be two-dimensional. What we learn here can be carried over to some extent to the higher dimensional case.

We start with a list of 50 5-simplices that form a 5-sphere. It is one example of Shemer's list from Bokowski and Shemer, 1987.

A 5-sphere of Shemer

```
facetsS::[[Int]]
facetsS = [[1,2,3,4,6,7],[1,2,3,4,5,6],[1,2,3,4,5,8],
  [1,3,4,5,6,9],[1,2,4,5,6,9],[2,3,4,5,6,10],[1,2,3,5,6,10],
  [1,2,3,4,7,10],[1,3,4,5,8,9],[1,2,4,5,8,9],[1,2,3,4,8,9],
  [2,3,4,5,8,10],[1,2,3,5,8,10],[1,2,3,4,9,10],[1,3,4,6,7,9],
  [1,2,4,6,7,9],[2,3,4,6,7,10],[1,2,3,6,7,10],[1,3,4,7,9,10],
  [1,2,4,7,9,10],[2,3,4,8,9,10],[1,2,3,8,9,10],[1,2,5,6,7,8],
  [1,3,5,6,7,9],[1,3,5,6,7,10],[1,2,5,6,7,10],[1,2,5,6,8,9],
  [1,2,5,7,8,10],[1,3,5,7,9,10],[1,3,5,8,9,10],[1,2,6,7,8,9],
  [1,2,7,8,9,10],[2,4,5,6,8,9],[2,4,5,6,8,10],[2,4,6,7,9,10],
  [2,4,6,8,9,10],[1,5,6,7,8,9],[2,5,6,7,8,10],[1,5,7,8,9,10],
  [2,6,7,8,9,10],[3,4,5,6,7,8],[3,4,5,6,7,9],[3,4,5,6,8,10],
  [3,4,5,7,8,9],[3,4,6,7,8,10],[3,4,7,8,9,10],[3,5,6,7,8,10],
  [3,5,7,8,9,10],[4,5,6,7,8,9],[4,6,7,8,9,10]]
```

We assume that there exists a 6-polytope the face lattice of which has these 5-simplices as facets. For such a simplicial polytope we can assume the vertices to be in general position. This implies that all hyperplanes that are spanned by the vertices of a subfacet and an additional vertex define a sequence of oriented simplices. When using function `facets2Chi`, we can determine a unique chirotope, a matroid polytope.

```
res1::[Int]
res1 = [1,1,1,1,1,1,1,-1,-1,-1,1,1,1,1,-1,-1,-1,-1,-1,-1,-1,-1,
  -1,1,-1,1,1,-1,1,1,-1,-1,1,1,-1,-1,-1,-1,1,-1,1,1,-1,1,1,-1,
  -1,1,1,-1,-1,-1,-1,-1,-1,-1,-1,-1,1,-1,1,-1,-1,1,-1,-1,1,1,
  -1,-1,1,1,-1,1,1,-1,-1,1,-1,1,1,-1,-1,1,1,-1,1,-1,1,-1,-1,1,
  -1,-1,1,1,-1,-1,1,1,-1,1,1,-1,-1,1,-1,1,1,-1,-1,1,1,-1,-1,1,
  1,1,1,-1,-1]
--res2 = dC (zip (tuples 7 10) res1)
```

This chirotope can be realized if and only if we can find coordinates for the dual rank 3 chirotope which has again ten elements. When we use the last three rows as a unit matrix, we see that the determinants of the submatrices of rows $[i, 9, 10]$ form the entries in the first column of rows $1, \ldots, 7$. We reorient those elements i for which $[i, 9, 10]$ is negative. Reorientation does not change the realization property of the chirotope. This realization property does not change

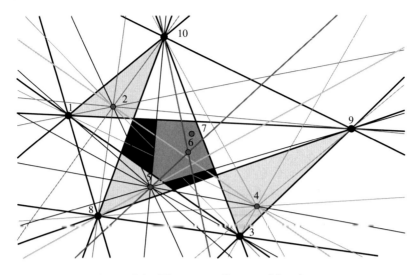

Figure 8.1   Planar coordinates of 6-polytope

when we multiply the (10×3)-matrix by the matrix

$$\begin{pmatrix} 1 & 0 & 0 \\ 1 & 1 & 0 \\ 1 & 0 & 1 \end{pmatrix}.$$

from the right and when we multiply the ith row with the reciprocal absolute value of the determinant $[i, 9, 10]$. The resulting realization problem is a problem in the affine plane. The convex hull of the chirotope tells us that we have a pentagon as its convex hull.

The picture in Figure 8.1 shows even the arguments towards the realization of this sphere. We have to find an additional five points in the pentagon such that all signs for the triangle orientations are correct. One point after another can be inserted; The corresponding bounds for the remaining points can be determined. Inserting one point at a time allows us to have control over the extensions. A dynamical software like Cinderella is again of much help in finding the final realization. This example was typical for many decisions that were obtained for the results in the article by Bokowski and Shemer, 1987.

8.2   Linear programming can help

A given chirotope, 10 elements, rank 4

```
chirotope::[Int]
chirotope
  = [ 1, 1, 1, 1, 1, 1, 1,-1,-1,-1,-1,-1,-1,-1,-1,-1,-1, 1,-1, 1,-1,-1,-1,-1,
     1,-1,-1,-1,-1,-1,-1,-1,-1,-1,-1,-1,-1,-1, 1,-1, 1,-1,-1,-1,-1, 1,-1,-1, 1,
     1,-1, 1,-1,-1,-1,-1,-1,-1,-1,-1, 1,-1,-1, 1,-1, 1,-1,-1,-1,-1, 1,-1,-1, 1,
```

```
    1,  1,-1,  1,  1,-1,-1,-1,  1,-1,-1,-1,-1,-1,-1,  1,  1,  1,  1,-1,  1,-1,-1,-1,-1,
   -1,-1,  1,-1,-1,-1,-1,-1,-1,-1,-1,-1,  1,-1,-1,  1,-1,  1,-1,-1,-1,  1,-1,  1,  1,
    1,  1,-1,  1,  1,  1,-1,  1,  1,-1,-1,  1,-1,-1,-1,-1,-1,-1,-1,-1,-1,  1,-1,-1,
    1,-1,  1,-1,-1,-1,  1,-1,  1,  1,  1,  1,-1,  1,  1,  1,-1,  1,  1,-1,-1,-1,
    1,  1,-1,  1,  1,-1,  1,-1,  1,  1,-1,  1,-1,-1,  1,-1,-1,-1,  1,-1,  1,-1,  1,
   -1,  1,-1,-1,-1,-1,  1,  1,  1,-1]
```

We use an oriented matroid represented as a chirotope that occurred during an investigation of triangulated oriented 2-manifolds of genus 3 of Bokowski and Lutz. It forms the input (chirotope) for our following Haskell program. Instead of looking for a matrix of our chirotope directly, we try to realize the dual uniform oriented matroid. Duality was discussed in Section 2.1 and was also used in the last section.

We have obtained the following dual chirotope

The dual chirotope

```
dCr::[Int]   -- result of   dC 4 10 chirotope
dCr
 = [-1,-1,  1,-1,-1,-1,  1,-1,-1,-1,-1,-1,  1,  1,  1,  1,-1,  1,  1,  1,-1,-1,  1,  1,  1,
      1,-1,-1,  1,  1,  1,-1,-1,-1,  1,-1,  1,-1,-1,-1,  1,  1,  1,  1,  1,-1,  1,  1,-1,  1,
     -1,  1,  1,  1,  1,-1,  1,  1,-1,-1,-1,  1,  1,  1,-1,-1,  1,-1,  1,-1,  1,-1,  1,  1,  1,
     -1,-1,-1,-1,-1,  1,-1,-1,  1,-1,  1,-1,-1,-1,-1,  1,-1,-1,  1,  1,  1,-1,-1,-1,  1,
      1,-1,  1,-1,  1,-1,  1,  1,-1,  1,-1,  1,-1,  1,  1,-1,-1,  1,-1,  1,  1,-1,  1,-1,  1,
     -1,  1,  1,-1,  1,  1,-1,-1,  1,  1,-1,  1,  1,-1,  1,  1,-1,-1,-1,-1,-1,  1,  1,-1,
      1,-1,  1,-1,  1,-1,-1,-1,-1,  1,  1,  1,-1,-1,  1,  1,-1,  1,-1,-1,-1,-1,  1,-1,  1,-1,
      1,-1,  1,-1,  1,-1,  1,-1,  1,  1,-1,  1,-1,  1,  1,  1,  1,-1,  1,-1,  1,-1,  1,-1,  1,
     -1,  1,-1,  1,-1,  1,-1,  1,-1,  1]
```

A matrix realization of this dual oriented matroid can be assumed to be of the following form

$$
\begin{array}{cccccc}
1 & 2 & 3 & 4 & 5 & 6 \\
\end{array}
$$

$$
\left(
\begin{array}{cccccc}
-1 & -e & -i & -m & -q & -u \\
1 & -f & -j & -n & -r & -v \\
-1 & -g & -k & -o & -s & -w \\
-1 & -1 & -1 & -1 & -1 & -1 \\
1 & 0 & 0 & 0 & 0 & 0 \\
0 & 1 & 0 & 0 & 0 & 0 \\
0 & 0 & 1 & 0 & 0 & 0 \\
0 & 0 & 0 & 1 & 0 & 0 \\
0 & 0 & 0 & 0 & 1 & 0 \\
0 & 0 & 0 & 0 & 0 & 1 \\
\end{array}
\right)
\begin{array}{c}
1 \\ 2 \\ 3 \\ 4 \\ 5 \\ 6 \\ 7 \\ 8 \\ 9 \\ 10 \\
\end{array}
$$

Before we have chosen variables, we have multiplied our matrix by the inverse of the submatrix of rows 5,6,7,8,9,10. In addition, we have multiplied row 1 with the absolute value of the inverse of submatrix of rows 1,6,7,8,9,10, we have multiplied row 2 with the absolute value of the inverse of submatrix of rows 5,1,7,8,9,10, etc. This has led to the values in the first column. Further on, we have multiplied the second column with the absolute value of the inverse of submatrix of rows 4,5,7,8,9,10, we have multiplied the third column with the absolute value of the inverse of submatrix of rows 4,5,6,8,9,10, etc. This has led to the values in the fourth row. All of these assumptions do not change the chirotope. We can now formulate that we look for 15 variables such that the above matrix has the dual of our original chirotope as its chirotope.

A three-term Grassmann–Plücker relation in this rank 6 case is of the following form where "−−" denotes "a, b, c, d."

$$\{a, b, c, d | e, f, g, h\} = \{-- | e, f, g, h\}$$
$$- [--, e, f] * [--, g, h] \quad [\quad, c, g] \, {}_{\text{II}} [\quad, f, h]$$
$$+ [--, e, h] * [--, f, g]$$
$$= 0.$$

The signs of these determinants are known from our dual chirotope. The uniform chirotope property ensures that all signs are either $+1$ or -1 and two equal signs of the three monomials A, B, C in the relation always differ from the third. This implies in the realizable case two inequalities for the absolute values $|A|, |B|, |C|$ of our three monomials A, B, C. We have six possible cases:

$(\text{sign} A = \text{sign} B = 1)$ or $(\text{sign} A = \text{sign} B = -1)$ imply $|A| < |C|$ and $|B| < |C|$;
$(\text{sign} A = \text{sign} C = 1)$ or $(\text{sign} A = \text{sign} C = -1)$ imply $|A| < |B|$ and $|C| < |B|$;
$(\text{sign} B = \text{sign} C = 1)$ or $(\text{sign} B = \text{sign} C = -1)$ imply $|B| < |A|$ and $|C| < |A|$.

When we take the logarithms on both sides of these resulting inequalities, we see that in the realizable case there must be a linear program solution for the new variables (logarithms of the absolute values of the brackets). This idea was first mentioned by M. Perles during Bokowski's investigation of Altshuler's spheres. It was tested much later and applied in Bokowski and Richter-Gebert, 1990.

The case *no solution exists* can be tested by a linear programming algorithm, there exists a so-called bi-quadratic final polynomial, see also Page 213.

Hence, it is useful to apply a linear programming algorithm to test whether we can find a bi-quadratic final polynomial.

However, we can do more with the solution of the linear inequality system. This following idea has still to be tested further. Here we sketch the method. A solution of our realizability problem is also a solution of the linear inequality system. Although the reverse it not true, we can use in the realizable case an interior solution point of the linear program as a good starting point for finding a realization. We can check how many signs for our variables are violated and we can try to change the linear programming solution into a solution of our original problem.

A small number of signs of the chirotope determine the remaining ones. We can determine such a reduced system and see how many inequalities are violated.

The following Haskell code has been used.

```
gpRelations::[([Int],[Int])]   --  { a,b,c,d | v,w,x,y }
gpRelations    -- we prepare all Grassmann--Pluecker relations
 = [([a,b,c,d],[v,w,x,y])
    | [a,b,c,d]<-tuplesL 4 [1..10],
      [v,w,x,y]<-tuplesL 4 ([1..10]\\[a,b,c,d])]

-- take 2 gpRelations
--[(([1,2,3,4],[5,6,7,8]),([1,2,3,4],[5,6,7,9])]
```

```
gpRel2Products::([Int],[Int])->[([Int],Int)]   -- ordered
gpRel2Products  gprel@([a,b,c,d],[v,w,x,y])
 = [ norm ([a,b,c,d,v,w],1), norm ([a,b,c,d,x,y],1),
     norm ([a,b,c,d,v,x],1), norm ([a,b,c,d,y,w],1),
     norm ([a,b,c,d,v,y],1), norm ([a,b,c,d,w,x],1) ]

--take 2 (map gpRel2Products gpRelations)
--[ [(([1,2,3,4,5,6], 1),([1,2,3,4,7,8],1),([1,2,3,4,5,7],1),
--    ([1,2,3,4,6,8],-1),([1,2,3,4,5,8],1),([1,2,3,4,6,7],1)],
--   [(([1,2,3,4,5,6], 1),([1,2,3,4,7,9],1),([1,2,3,4,5,7],1),
--    ([1,2,3,4,6,9],-1),([1,2,3,4,5,9],1),([1,2,3,4,6,7],1)]]

gpRelProducts = (map gpRel2Products gpRelations)
```

```
bracket2Index::([Int],Int)->Int -- indices for all brackets
bracket2Index  br = head (elemIndices (fst br) (tuples 6 10))

modifiedSign::([Int],Int)->Int   -- dC inserted
modifiedSign br = (dC!!(bracket2Index br))*(snd br)

g::[(Int,Int)]->[((Int,Int),(Int,Int))]
g list@[(a,sa),(b,sb),(c,sc),(d,sd),(e,se),(f,sf)]
 | sa*sb ==  1 && sc*sd ==  1 = [((a,b),(e,f)),((c,d),(e,f))]
 | sa*sb ==  1 && se*sf ==  1 = [((a,b),(c,d)),((e,f),(c,d))]
 | sc*sd ==  1 && se*sf ==  1 = [((c,d),(a,b)),((e,f),(a,b))]
 | sa*sb == -1 && sc*sd == -1 = [((a,b),(e,f)),((c,d),(e,f))]
 | sa*sb == -1 && se*sf == -1 = [((a,b),(c,d)),((e,f),(c,d))]
 | sc*sd == -1 && se*sf == -1 = [((c,d),(a,b)),((e,f),(a,b))]

--((0,9),(2,5)),((1,6),(2,5)) 0 * 9 < 2 * 5 and 1 * 6 < 2 * 5
```

```
main=print(concat(map g (map part) gpRelProducts)))
    where part=map(\br->(bracket2Index br,modifiedSign br))
{-
[((0,9),(2,5)),((1,6),(2,5)),((0,10),(1,7)),
 ((3,5),(1,7)),((0,11),(1,8)),((4,5),(1,8)),.etc.,
 ((196,209),(201,208)),((202,207),(201,208))] -}
```

For the resulting linear inequality system we have found the following solution. Only 15 variable entries of our matrix (among those 210 variables) determine all the other signs of determinants that we want to find. In other words, we use the projection of our solution point onto our 15-dimensional parameter space and we check the number of violated inequalities for our required determinants. It has turned out that only a few inequalities were violated and a change of the projected point (one coordinate at a time) led finally to the following solution.

x0	72.654425245085847	x70	13.320253547118034	x140	0.000045399929762
x1	16.651494963610141	x71	94.970982730440326	x141	135.736336253693224
x2	331.481300736422554	x72	38.889111444910817	x142	39354.161331196097308
x3	53.155113021440222	x73	19.906999429230556	x143	0.016454720211215
x4	40.664566040939519	x74	16.651494963610141	x144	2584.126709798547836
x5	2.335472649213012	x75	0.914584226723249	x145	0.122675294053084
x6	42.521082000062776	x76	0.874652538802888	x146	4222.633649805529785
x7	885.112096766640093	x77	35.567368416741161	x147	39354.161331196097308
x8	5.963838396263028	x78	0.022490935896088	x148	231.928376031711082
x9	0.000045399929762	x79	1.562721058404001	x149	4222.633649805529785
x10	148.413159102576572	x80	1.634066069776326	x150	0.000270757844305
x11	0.000045399929762	x81	129.809942506527221	x151	518.012824668341864
x12	3377.867931673533349	x82	0.183337126433589	x152	51442.193504358758219
x13	0.001614756027956	x83	253.588861482623173	x153	0.409484126322319
x14	303.167573351583542	x84	0.000045399929762	x154	7888.918245966450741
x15	1581.408840625033008	x85	86.858964017692230	x155	0.299584998606369
x16	289.930981549963747	x86	155.188871356346596	x156	50.834303884147843
x17	38.889111444910817	x87	0.000054275990104	x157	0.117319171122215
x18	32.529354595417075	x88	7.455325431147260	x158	0.000045399929762
x19	3861.954185410319496	x89	0.117319171122215	x159	0.286504796860190
x20	774.165519337122760	x90	647.561843364206311	x160	4.770733181967603
x21	566.391593338632674	x91	541.662384215076031	x161	619.288600065196192
x22	0.098133292195410	x92	141.933281699235124	x162	518.012824668341864
x23	0.017205950425851	x93	48.614828619708391	x163	118.722127550431310
x24	0.000755976701788	x94	0.052526966165786	x164	677.125816628096118
x25	473.766365314750544	x95	1.634066069776326	x165	0.000074186482570
x26	18424.352979575673089	x96	566.391593338632674	x166	242.516900322328695
x27	94.970982730440326	x97	0.428178852934532	x167	541.662384215076031
x28	3230.386601754571075	x98	23.798984221554395	x168	12.738678291730114
x29	0.005154644634350	x99	0.000578335099711	x169	19.037840427406088
x30	253.588861482623173	x100	1156.983423555662966	x170	31.109088150967658
x31	41150.855677666746487	x101	0.023517745856009	x171	885.112096766640093
x32	0.020569855502784	x102	473.766365314750544	x172	0.731615628946642
x33	4038.268944307595575	x103	0.250592140435207	x173	396.288665904250365
x34	0.117319171122215	x104	2.335472649213012	x174	0.669124116935304
x35	9.745305676603497	x105	0.191707271113876	x175	72.654425245085847
x36	50.834303884147843	x106	0.175332448763563	x176	967.775365584676251
x37	19.906999429230556	x107	0.000045399929762	x177	66.448592261079170
x38	13.928381598082510	x108	0.262032771757337	x178	809.509573917844477
x39	14.564271907856952	x109	0.122675294053084	x179	63.547375589457630
x40	0.874652538802888	x110	0.153354966844928	x180	647.561843364206311
x41	0.699672535376066	x111	0.200459529598193	x181	1.000000000000000
x42	14.564271907856952	x112	0.000045399929762	x182	1106.468399374343562
x43	1.634066069776326	x113	0.011010250240849	x183	967.775365584676251
x44	0.128275959776909	x114	0.048040335405419	x184	177.429154559510550
x45	1.562721058404001	x115	0.000045399929762	x185	1156.983423555662966
x46	103.840607158846794	x116	0.000442436193232	x186	0.000121225622489
x47	0.140256035729623	x117	0.000045399929762	x187	378.986278522155430
x48	193.999794541823007	x118	0.200459529598193	x188	1058.158804578495847
x49	0.098133292195410	x119	0.018812865652203	x189	0.167677246356408
x50	72.654425245085847	x120	1.045654345083818	x190	48.614828619708391
x51	141.933281699235124	x121	0.914584226723249	x191	0.468167767377395
x52	0.000045399929762	x122	0.956338970618289	x192	1209.804796052536403
x53	5.454432604310696	x123	0.078501079711637	x193	0.000258936266975
x54	10.190221663089968	x124	0.836464306098034	x194	453.081218636183849
x55	317.008503994966020	x125	1.000000000000000	x195	0.000045399929762
x56	253.588861482623173	x126	740.364698051824348	x196	0.639909467286032
x57	90.824457040030509	x127	0.000045399929762	x197	1265.037569588432689
x58	21.766173709178108	x128	1106.468399374343562	x198	1.143311150012405
x59	3.490342957461841	x129	0.000045399929762	x199	846.467155105515303
x60	0.175332448763563	x130	3532.082431753260153	x200	1.045654345083818

x61	242.516900322328695	x131	1265.037569588432689	x201	10.190221663089968
x62	29.750832059884218	x132	378.986278522155430	x202	1.000000000000000
x63	2.042727061511597	x133	4616.998049285642082	x203	1.000000000000000
x64	0.000045399929762	x134	0.000323693233780	x204	1.000000000000000
x65	809.509573917844477	x135	677.125816628096118	x205	1.000000000000000
x66	2.042727061511597	x136	473.766365314750544	x206	1.000000000000000
x67	433.299250935064549	x137	10782.872773405180851	x207	1.000000000000000
x68	19.906999429230556	x138	58.119428177447972	x208	1.000000000000000
x69	193.999794541823007	x139	453.081218636183849	x209	1.000000000000000

Finally, we can forget all the data that was used during the realization process. We have found coordinates for a corresponding embedded 2-manifold in 3-space with ten vertices.

8.3   The final polynomial method

In the study of projective incidence theorems, we have encountered a polynomial that was decisive in proving non-realizability of oriented matroids. We best understand searching and finding such a polynomial by working on a particular example. The polynomial that finally shows up will be our final polynomial. Such a detailed description has been given in Bokowski, 1991. We find a more general description in Bokowski, 2001.

8.4   Solvability sequences

In papers of Bokowski and Sturmfels, we find the notion *solvability sequence* where we can solve the non-linear system of algebraic inequalities up to the end step-by-step without running into trouble. These cases occur especially if we have a small number of elements and a small rank as in the cases that were relevant in those articles.

8.5   Combinatorial arguments

In Richter-Gebert, 1989 and 1991, he investigated cases where we can find reductions already on a combinatorial level. Such reduction methods are, as with all our heuristic methods, very good provided they succeed. We can always test them. If all heuristic methods fail, we only have the theoretical result that in principle we have an algorithm. However, as we have mentioned before at the beginning of this chapter, general algorithmic versions from real algebraic geometry are far from being applicable for our problem sizes.

8.6   Non-realizable oriented matroids

Whenever we can argue that the oriented matroid of the realized geometrical object must have a non-realizable oriented matroid as one of its minors, we have shown that the object cannot be embedded. So the non-realizable oriented matroids can play a decisive role. Compare with Page 285.

8.7   Finding a polytope with a given face lattice

We start with the following simplicial face lattice, that we have encountered at the end of Chapter 4. The interesting aspect of this 3-sphere was the symmetry of the face lattice. The combinatorial symmetry cannot be realized by an affine symmetry. The realization space is not contractable and even not connected. We assume now that we do not know whether this sphere is polytopal.

$$
\begin{array}{llll}
[[1,2,3, \ 4], & [1,2,3, \ 7], & [1,2,4, \ 8], & [1,2,6, \ 7], \\
[1,2,6, \ 8], & [1,3,4, \ 7], & [1,4,5, \ 6], & [1,4,5, \ 8], \\
[1,4,6, \ 7], & [1,5,6, \ 8], & [2,3,4, \ 8], & [2,3,7,10], \\
[2,3,8, \ 9], & [2,3,9,10], & [2,6,7, \ 9], & [2,6,8, \ 9], \\
[2,7,9,10], & [3,4,5, \ 7], & [3,4,5, \ 8], & [3,5,7,10], \\
[3,5,8,10], & [3,8,9,10], & [4,5,6, \ 7], & [5,6,7, \ 9], \\
[5,6,8,10], & [5,6,9,10], & [5,7,9,10], & [6,8,9,10]]
\end{array}
$$

We describe a method for finding coordinates. First of all, we have to determine admissable matroid polytopes. There are several of them and we pick a particular one.

```
[((1,2,3,4, 5],-1),((1,2,3,4, 6],-1),((1,2,3,4, 7],-1),((1,2,3,4, 8],-1),((1,2,3,4, 9],-1),((1,2,3,4,10],-1),
((1,2,3,5, 6],-1),((1,2,3,5, 7],-1),((1,2,3,5, 8], 1),((1,2,3,5, 9], 1),((1,2,3,5,10], 1),((1,2,3,6, 7],-1),
((1,2,3,6, 8], 1),((1,2,3,6, 9], 1),((1,2,3,6,10], 1),((1,2,3,7, 8], 1),((1,2,3,7, 9], 1),((1,2,3,7,10], 1),
((1,2,3,8, 9],-1),((1,2,3,8,10],-1),((1,2,3,9,10],-1),((1,2,4,5, 6],-1),((1,2,4,5, 7],-1),((1,2,4,5, 8], 1),
((1,2,4,5, 9], 1),((1,2,4,5,10], 1),((1,2,4,6, 7],-1),((1,2,4,6, 8], 1),((1,2,4,6, 9], 1),((1,2,4,6,10], 1),
((1,2,4,7, 8], 1),((1,2,4,7, 9], 1),((1,2,4,7,10], 1),((1,2,4,8, 9],-1),((1,2,4,8,10],-1),((1,2,4,9,10],-1),
((1,2,5,6, 7],-1),((1,2,5,6, 8], 1),((1,2,5,6, 9], 1),((1,2,5,6,10], 1),((1,2,5,7, 8],-1),((1,2,5,7, 9],-1),
((1,2,5,7,10],-1),((1,2,5,8, 9],-1),((1,2,5,8,10],-1),((1,2,5,9,10],-1),((1,2,6,7, 8],-1),((1,2,6,7, 9],-1),
((1,2,6,7,10],-1),((1,2,6,8, 9], 1),((1,2,6,8,10], 1),((1,2,6,9,10], 1),((1,2,7,8, 9],-1),((1,2,7,8,10],-1),
((1,2,7,9,10], 1),((1,2,8,9,10],-1),((1,3,4,5, 6],-1),((1,3,4,5, 7],-1),((1,3,4,5, 8], 1),((1,3,4,5, 9], 1),
((1,3,4,5,10], 1),((1,3,4,6, 7],-1),((1,3,4,6, 8], 1),((1,3,4,6, 9], 1),((1,3,4,6,10], 1),((1,3,4,7, 8], 1),
((1,3,4,7, 9], 1),((1,3,4,7,10], 1),((1,3,4,8, 9],-1),((1,3,4,8,10],-1),((1,3,4,9,10],-1),((1,3,5,6, 7],-1),
((1,3,5,6, 8], 1),((1,3,5,6, 9], 1),((1,3,5,6,10], 1),((1,3,5,7, 8], 1),((1,3,5,7, 9], 1),((1,3,5,7,10], 1),
((1,3,5,8, 9],-1),((1,3,5,8,10],-1),((1,3,5,9,10],-1),((1,3,6,7, 8],-1),((1,3,6,7, 9],-1),((1,3,6,7,10],-1),
((1,3,6,8, 9], 1),((1,3,6,8,10], 1),((1,3,6,9,10], 1),((1,3,7,8, 9],-1),((1,3,7,8,10],-1),((1,3,7,9,10], 1),
((1,3,8,9,10],-1),((1,4,5,6, 7], 1),((1,4,5,6, 8], 1),((1,4,5,6, 9], 1),((1,4,5,6,10], 1),((1,4,5,7, 8], 1),
((1,4,5,7, 9], 1),((1,4,5,7,10], 1),((1,4,5,8, 9],-1),((1,4,5,8,10],-1),((1,4,5,9,10], 1),((1,4,6,7, 8], 1),
((1,4,6,7, 9], 1),((1,4,6,7,10], 1),((1,4,6,8, 9], 1),((1,4,6,8,10], 1),((1,4,6,9,10], 1),((1,4,7,8, 9], 1),
((1,4,7,8,10], 1),((1,4,7,9,10], 1),((1,4,8,9,10],-1),((1,5,6,7, 8], 1),((1,5,6,7, 9], 1),((1,5,6,7,10], 1),
((1,5,6,8, 9], 1),((1,5,6,8,10], 1),((1,5,6,9,10], 1),((1,5,7,8, 9], 1),((1,5,7,8,10], 1),((1,5,7,9,10], 1),
((1,5,8,9,10],-1),((1,6,7,8, 9], 1),((1,6,7,8,10],-1),((1,6,7,9,10],-1),((1,6,8,9,10], 1),((1,7,8,9,10],-1),
((2,3,4,5, 6],-1),((2,3,4,5, 7],-1),((2,3,4,5, 8], 1),((2,3,4,5, 9], 1),((2,3,4,5,10], 1),((2,3,4,6, 7],-1),
((2,3,4,6, 8], 1),((2,3,4,6, 9], 1),((2,3,4,6,10], 1),((2,3,4,7, 8], 1),((2,3,4,7, 9], 1),((2,3,4,7,10], 1),
((2,3,4,8, 9],-1),((2,3,4,8,10],-1),((2,3,4,9,10],-1),((2,3,5,6, 7],-1),((2,3,5,6, 8],-1),((2,3,5,6, 9],-1),
((2,3,5,6,10],-1),((2,3,5,7, 8],-1),((2,3,5,7, 9], 1),((2,3,5,7,10], 1),((2,3,5,8, 9],-1),((2,3,5,8,10],-1),
((2,3,5,9,10],-1),((2,3,6,7, 8],-1),((2,3,6,7, 9], 1),((2,3,6,7,10], 1),((2,3,6,8, 9],-1),((2,3,6,8,10],-1),
((2,3,6,9,10],-1),((2,3,7,8, 9],-1),((2,3,7,8,10],-1),((2,3,7,9,10],-1),((2,3,8,9,10],-1),((2,4,5,6, 7], 1),
((2,4,5,6, 8],-1),((2,4,5,6, 9],-1),((2,4,5,6,10],-1),((2,4,5,7, 8], 1),((2,4,5,7, 9], 1),((2,4,5,7,10], 1),
((2,4,5,8, 9],-1),((2,4,5,8,10],-1),((2,4,5,9,10],-1),((2,4,6,7, 8], 1),((2,4,6,7, 9], 1),((2,4,6,7,10], 1),
((2,4,6,8, 9],-1),((2,4,6,8,10],-1),((2,4,6,9,10],-1),((2,4,7,8, 9], 1),((2,4,7,8,10], 1),((2,4,7,9,10],-1),
((2,4,8,9,10], 1),((2,5,6,7, 8],-1),((2,5,6,7, 9], 1),((2,5,6,7,10],-1),((2,5,6,8, 9],-1),((2,5,6,8,10],-1),
((2,5,6,9,10],-1),((2,5,7,8, 9], 1),((2,5,7,8,10], 1),((2,5,7,9,10], 1),((2,5,8,9,10], 1),((2,6,7,8, 9], 1),
((2,6,7,8,10],-1),((2,6,7,9,10],-1),((2,6,8,9,10], 1),((2,7,8,9,10], 1),((3,4,5,6, 7], 1),((3,4,5,6, 8],-1),
((3,4,5,6, 9],-1),((3,4,5,6,10],-1),((3,4,5,7, 8],-1),((3,4,5,7, 9], 1),((3,4,5,7,10], 1),((3,4,5,8, 9],-1),
((3,4,5,8,10],-1),((3,4,5,9,10],-1),((3,4,6,7, 8],-1),((3,4,6,7, 9], 1),((3,4,6,7,10], 1),((3,4,6,8, 9],-1),
((3,4,6,8,10],-1),((3,4,6,9,10],-1),((3,4,7,8, 9],-1),((3,4,7,8,10],-1),((3,4,7,9,10],-1),((3,4,8,9,10],-1),
((3,5,6,7, 8], 1),((3,5,6,7, 9], 1),((3,5,6,7,10], 1),((3,5,6,8, 9],-1),((3,5,6,8,10],-1),((3,5,6,9,10], 1),
((3,5,7,8, 9],-1),((3,5,7,8,10],-1),((3,5,7,9,10],-1),((3,5,8,9,10], 1),((3,6,7,8, 9], 1),((3,6,7,8,10],-1),
((3,6,7,9,10],-1),((3,6,8,9,10], 1),((3,7,8,9,10], 1),((4,5,6,7, 8], 1),((4,5,6,7, 9], 1),((4,5,6,7,10], 1),
((4,5,6,8, 9],-1),((4,5,6,8,10],-1),((4,5,6,9,10], 1),((4,5,7,8, 9], 1),((4,5,7,8,10], 1),((4,5,7,9,10],-1),
((4,5,8,9,10], 1),((4,6,7,8, 9],-1),((4,6,7,8,10],-1),((4,6,7,9,10],-1),((4,6,8,9,10],-1),((4,7,8,9,10],-1),
((5,6,7,8, 9],-1),((5,6,7,8,10],-1),((5,6,7,9,10], 1),((5,6,8,9,10], 1),((5,7,8,9,10],-1),((6,7,8,9,10],-1)]
```

We determine a small system of determinants such that their proper signs imply the remaining signs via Grassmann–Plücker relations (a reduced system). The choice of the unit matrix was taken to support this idea of getting a small reduced system.

$$M := \begin{pmatrix} 1 & 0 & 0 & 0 & 0 \\ -a & -b & -c & d & e \\ 0 & 1 & 0 & 0 & 0 \\ f & g & h & -i & -j \\ 0 & 0 & 1 & 0 & 0 \\ k & -l & -m & n & o \\ 0 & 0 & 0 & 1 & 0 \\ 0 & 0 & 0 & 0 & 1 \\ -p & -q & -r & s & t \\ -u & -v & w & x & y \end{pmatrix}.$$

We use in particular the variables $a, b, c, \ldots, y$ with their signs as indicated. This implies: 25 signs have correct values.

These following inequalities follow from the reduced system:

```
cj < eh      em < co      qx < sv
dh < ci      eq < bt      mq < lr
do < en      nr < ms      fr < hp
es < dt      go < jl      kw < mu
ex < dy      nv < lx      gu < fv
```

```
ilw + imv + hlx < gnw + gmx + hnv
aho + efm + ehk < cjk + ajm + cfo
epw + eru + cpy < ary + ctu + atw
arx + csu + asw < dpw + dru + cpx
ags + bip + dfq < dgp + bfs + aiq
alt + bop + bkt < ekq + elp + aoq
ekv + elu + aov < aly + bou + bky
```

We can still apply a normalization. Without loss of generality we set $c = h = m = r = w = 1$ and $k = l = n = o = 1$. This leads to the following system of inequalities for positive variables.

```
q < 1 < s               es < dt
q < j < e < 1 < u       ex < dy
d < i                   gu < fv
d < e                   eq < bt
f < p                   qx < sv
v < x
```

```
  i +  iv +   x  <   g +  gx +    v
  a +  ef +   e  <   j +  aj +    f
 ep +  eu +  py  <  ay +  tu +   at
 ax +  su +  as  <  dp +  du +   px
 at +  bp +  bt  <  eq +  ep +   aq
 ev +  eu +  av  <  ay +  bu +   by
ags + bip + dfq  < dgp + bfs +  aiq
```

The non-linear system of inequalities is still linear in each variable. We check carefully for which variable we should solve. Solving for j, y, t, i, f in this sequence is a good way towards the final solution. In Bokowski, 1991 we find a more detailed sequence of inequalities. A corresponding computer algebra package such as maple supports very much the solution process.

For a large class of Altshuler spheres this method was applied and tested as an algorithmic approach for polytopal and non-polytopal spheres in Bokowski and Sturmfels, 1987. See also Bokowski and Sturmfels, 1986.

9

Computational synthetic geometry

Instead of extracting combinatorial properties from geometrical objects, *Computational Synthetic Geometry* deals with the reverse problem. Starting with given combinatorial properties of a questionable geometrical object, we try to determine such a geometrical object with the given properties. Sometimes we look for a proof that the questionable geometric object does not exist. Large problem classes can be summarized under this heading. Our Chapters 6, 7, and 8 were already contributions to this general problem class. The title *Computational Synthetic Geometry* was first used in Bokowski, 1987 and Sturmfels, 1987, before it was chosen as a book title, see Bokowski and Sturmfels, 1989b.

The theory of oriented matroids has proven useful in this context. Especially, when the problem size is small enough, it has turned out that even a complete overview about all realizations can be obtained. Above all, we have a tool that stresses the distinction between combinatorial and metrical properties of the object. Oriented matroids form an intermediate mathematical concept that has to exist in the realizable case, and sometimes does not which then implies that the geometrical object does not exist.

On the other hand, we cannot promise that our tools from oriented matroid theory will be successful for all problem types. When the realizability problem for oriented matroids remains, we know that the complexity of this remaining problem of solving non-linear polynomial equations and inequalities is double exponential in the number of variables. Even deciding whether a matroid can be oriented is very difficult, and finding all oriented matroids for a given problem might lie beyond what our computers can do for us.

So, very often, intuitive methods have also led to solutions. The competition between ingenious and intuitive guesses of how the object might look and computational approaches towards a systematic solution remains.

In this chapter we first discuss polyhedral tori with the smallest number of vertices and with the smallest number of facets. We than focus on the attempt to

217

generalize a celebrated theorem of Steinitz for characterizing combinatorial types of convex 3-polytopes.

There are two possible generalizations, we can look at non-convex polyhedra and we can try to find results in higher dimensions. We study various non-convex polyhedra, both realizable and non-realizable ones.

9.1 On Möbius's torus and Császár's polyhedron

The smallest triangulated torus, with respect to the number of vertices, was presented by Möbius in his "Preisarbeit" in Paris in 1861, see Möbius, 1886.

Identifying two opposite triangles of the triangular prism of Figure 9.1 leads to Mobius's torus. This abstract triangulated closed 2-manifold of genus one has an automorphism group of order 42.

The pottery model of Figure 9.2 and the drawing of Karl Heinrich Hofmann in Figure 9.3 reveals a symmetry of order 7.

In Figure 9.4 we have depicted a Möbius torus tiling of the plane that shows a symmetry of order 3. Reflecting the triangulation at any edge is also possible. However, the combinatorial symmetry of the triangulated torus is not flag transitive. A symmetry of order 2 is missing. As usual, we define *a flag* of a map as a triple consisting of a face, an edge incident with that face, and a vertex incident with that edge. When for any pair of flags there exists an element of the combinatorial automorphism group of the map that maps one flag onto the other, we define the map to be a *regular map*. A list of regular maps on surfaces with small genus can be found in Coxeter and Moser, 1980.

Although Möbius's torus and its dual, the so-called Heawood map, are not regular maps in our sense, they form cell-decompositions of surfaces that are of

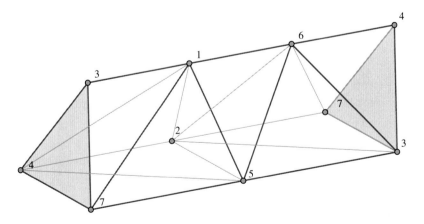

Figure 9.1   The torus of Möbius with seven vertices

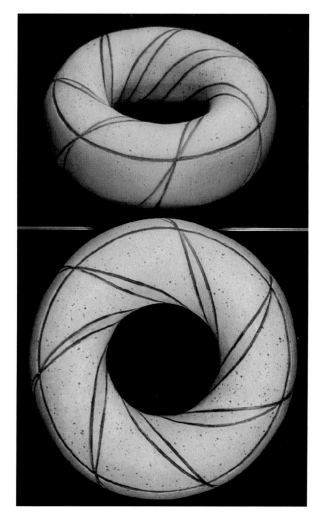

Figure 9.2 Pottery model of Möbius's torus revealing a symmetry of order 7

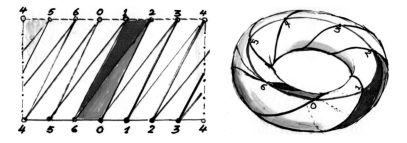

Figure 9.3 The torus of Möbius with seven vertices, combinatorial description (left), topological triangles (right); drawings of Karl Heinrich Hofmann, Darmstadt.

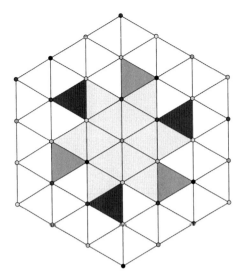

Figure 9.4   Möbius's torus tiling of the plane

interest as small examples of infinite series. Möbius's torus forms a triangular complete graph embedding such that no arcs intersect.

Can we embed Möbius's torus in 3-space with flat triangles and without self-intersections? Császár has answered this question in the affirmative, see Császár, 1949. We describe his realization by using Figure 9.5. We start with a square, we fold one of its diagonals and lift two vertices marked + a little. We use two additional vertices, marked ++, higher than those marked +, as indicated in Figure 9.5.

The projection shows four additional triangles and the third part of the drawing defines two additional triangles. The final step is to put a vertex marked ++++ on top over the center of the projection. We look at the star of this last vertex, that is, at all triangles incident with it. Those edges of these triangles that are not incident with this last vertex (the link of this vertex) form a hexagon. The upper and lower edges in Figure 9.5 of the original square are not part of this hexagon.

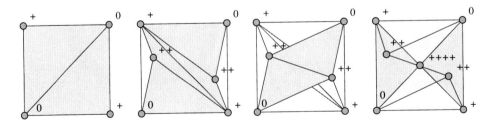

Figure 9.5   Realization of Möbius's torus due to Császár

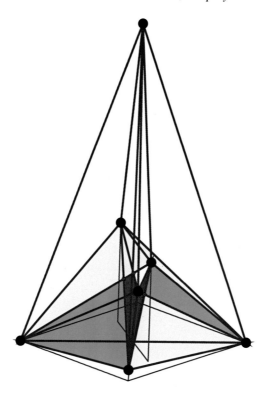

Figure 9.6   Intuitive solution of a realization of Möbius's torus, Császár's polyhedron

We have realized all 14 triangles without self-intersections and with a symmetry of order 2.

This realization is known as Császár's polyhedron. When M. Gardner was informed by D. Crowe about this remarkable polyhedron, he wrote an article in the *Scientific American* about this polyhedron and he made it well known not only to mathematicians, see Gardner, 1978. For a recent article about this polyhedron, see Lutz, 2001.

In Bokowski and Eggert, 1991, we find other realization types and a complete overview about all oriented matroid types that are all realizable. This article was a test for applying methods from the theory of oriented matroids in this context. Figure 9.7 shows one of those additional symmetrical realizations.

The convex hull is a tetrahedron. We see parts of the interior because of the missing slices. A film from 1986, produced by J. Bokowski and J. Richter-Gebert, supported by J. M. Wills and H. Simon from the video studio of the University of Siegen, shows altogether four symmetric realizations of Möbius's torus with seven vertices. Nowadays, the film is just a reminder of the old days of computer graphics when moving polyhedra in real time on the screen were not available on ordinary computers. You can see a version in `http:\\juergen.bokowski.de`.

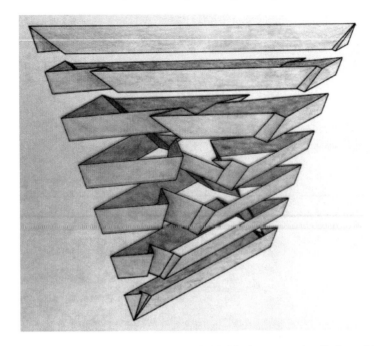

Figure 9.7   A symmetric realization of Möbius's torus, by Bokowski and Eggert, which was found via oriented matroid techniques

The realizations of all polyhedra were found by determining all oriented matroids that are compatible with the intersection conditions of triangles and edges. The symmetric ones can be detected on the combinatorial level. The inequality system for the coordinates was solved to arrive at the solution. A more detailed description was given in the article, Bokowski and Eggert, 1991. In the collected work of Möbius, we find a remark about a model of that torus with seven vertices due to Reinhardt, 1885. However, from the description it follows that this model had self-intersections, for a sketch of this model, see Bokowski, 1991.

9.2   On Heawood's map and Szilassi's polyhedron

We obtain the dual torus of Möbius's torus with seven vertices by inserting a vertex in each triangle and connecting vertices if and only if the triangles have a common edge.

The resulting combinatorial torus structure is known as Heawood's map, see Figure 9.8, where opposite sides of the rectangle have to be identified as well as Figure 9.9 where we see the map on the torus.

Former vertices of Möbius's torus are now hexagons, former triangles of Möbius's torus are now vertices, and former edges are edges again.

Figure 9.8   Heawood's map, computer graphic of Szilassi

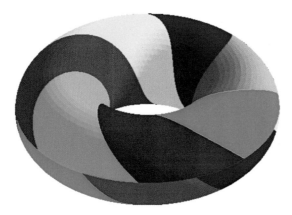

Figure 9.9   Heawood's map on the torus; computer graphic of Szilassi

Why do we mention Heawood's map? The chromatic number χ of a surface is the smallest number such that all maps on the surface can be colored with χ colors. Heawood proved in 1890 the following inequality about the chromatic number $\chi(S_g)$ on an orientable surface S_g of genus g:

$$\chi(g) \leq \lfloor \frac{1}{2}(7 + \sqrt{48g + 1}) \rfloor \text{ for } g \geq 1.$$

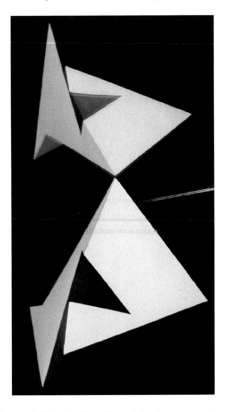

Figure 9.10   Szilassi's polyhedra on a mirror

Here $\lfloor . \rfloor$ denotes the floor function, that is, $\lfloor x \rfloor$ denotes the largest integer not greater than x. Thus Heawoods map shows for the torus that the chromatic number is 7.

For a first polyhedral realization of Heawood's map, see Szilassi, 1986. In Figure 9.10 we see a model of this toroidal polyhedron. See also `http://mathworld.wolfram.com/SzilassiPolyhedron.html`

Szilassi has asked whether there are, apart from his solution, essentially different symmetrical polyhedral realizations without self-intersections. With the following labeling `hexagonsSzilassi` and with the symmetry given by the Haskell function `symmetry` we find that the hexagon $(5, 12, 6, 10, 3, 9)$ has to be a point symmetric one. The rotational symmetry of order 2 for the whole polyhedron forces three edges to lie orthogonal to the axis of symmetry and parallel to the edges of the above hexagon.

We can now start an oriented matroid investigation in order to answer Szilassi's question. The matroid information can be used as our input, a required symmetry for the chirotope reduces essentially the computations. However, we can look at all realizations directly. After picking without loss of generality two opposite edges of the symmetric hexagon, there are two parameters left for the additional pair of points of this hexagon. The distance of one edge from this hexagon that remains fixed under the symmetry can be chosen arbitrarily. The oriented distances of the additional two edges that remain fixed under the symmetry define all the seven planes of the torus. A dynamical software such as Cinderella can help us to look at all these realizations. Such an investigation of the author has led to the conjecture that there is essentially only Szilassi's realization.

A deeper investigation has confirmed this conjecture when we require a symmetric realization, see Bokowski and Schewe, 2002.

Symmetry analysis for Szilassi's polyhedron

```
hexagonsSzilassi::[[Int]]
hexagonsSzilassi
 = [[1,  8,  2, 13,  6, 12], [2,  9,  3, 14,  7, 13],
    [3, 10,  4,  8,  1, 14], [4, 11,  5,  9,  2,  8],
    [5, 12,  6, 10,  3,  9], [6, 13,  7, 11,  4, 10],
                             [7, 14,  1, 12,  5, 11]]
symmetry::Int->Int; symmetry k = 15 - k


edges::[[Int]]->[[Int]]
edges [] = []
edges (hexagon:hexagons)
 = (edges6gon hexagon)++edges hexagons

edges6gon::[Int]->[[Int]]
edges6gon hexagon@[a,b,c,d,e,f]
 = [[a,b],[b,c],[c,d],[d,e],[e,f],[f,a]]

edgesSzilassi = nub (map sort (edges hexagonsSzilassi))
fixEdges      = [edge |edge<-edgesSzilassi,
                     head edge == symmetry (last edge)]
-- [[2,13],[1,14],[4,11]]
```

We have depicted Szilassi's realization in Figure 9.11 with a top view and a front elevation. The colors of the edges indicate the hexagons with which they are incident. This is just one picture of all realizations. The realization space is four-dimensional. By changing the coordinates of vertex 6 and the heights of the blue and the green edges that are fixed under the symmetry and by playing with

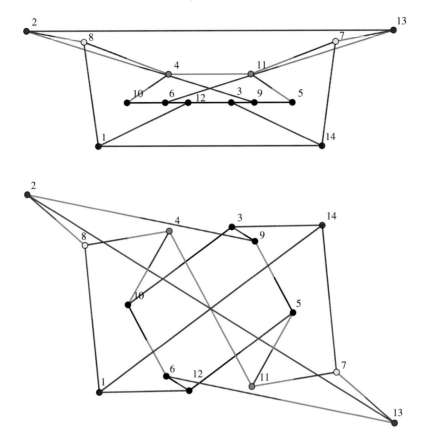

Figure 9.11   Szilassi's realization of a torus with seven hexagons

the software Cinderella, we see that we have in general self-intersections of edges belonging to the same hexagon. In Figure 9.12 this is not the case. However, there are self-intersections between hexagons.

9.2.1   A seven curve arrangement on a genus 3 surface

We can once more form the previous combinatorial incidence structure of Möbius's torus in a non-familiar way. Abstract points of the triangulated torus are now curves on a genus 3 surface. Abstract edges of the triangulated torus are now meets of curves. Abstract triangles of Möbius's torus with its vertices are now triangles on a genus 3 surface with its bounding curves. We see a corresponding pottery model in Figure 9.13: an oriented 2-manifold of genus 3 with seven closed curves that meet pairwise precisely once where they cross. We will discuss this observation from Bokowski and Pisanski, 2002, in a separate section.

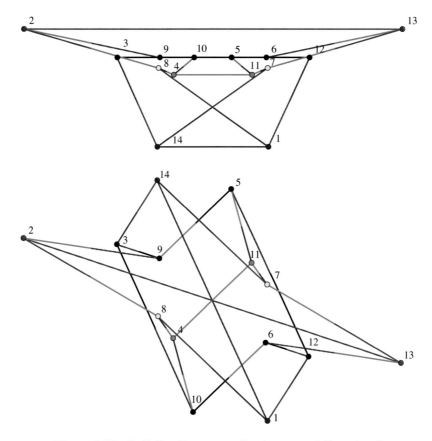

Figure 9.12 Is Szilassi's torus realization essentially unique?

9.3 Two general problems

Two general problems and corresponding partial contributions for solving them provide some important applications of the theory of oriented matroids. They both concern generalizations of the theorem of Steinitz, characterizing 2-spheres that occur as boundaries of 3-polytopes. If we replace 2-spheres with orientable triangulated closed 2-manifolds, we arrive at Problem 9.1. If we replace 2-spheres with higher dimensional spheres, we arrive at Problem 9.2.

Problem 9.1 *What is the smallest genus number g of an example of an oriented closed triangulated 2-manifold that allows no spatial embedding with flat triangles and without self-intersections?*

Problem 9.2 *Characterize, among all combinatorial $(d-1)$-spheres, the d-polytopal ones.*

Figure 9.13   Seven curve arrangement on an orientable surface of genus 3, pottery model

Problem 9.1 contains in particular the Steinitz problem for tori. A claim of three authors that such a solution was found has not been so far confirmed.

Problem 9.2 is known as the Steinitz problem for spheres.

Both problems very often lead to a subproblem that is difficult and fundamental in its own right, a main problem in the theory of oriented matroids that was the topic of Chapter 8.

Within this chapter we look in particular at the Steinitz problem for Tori.

We discuss orientable polyhedra with small genus. We discuss some known non-realizable examples. We mention some pinched 2-spheres without diagonals, immersions of 2-manifolds, and finally, we look at some polyhedral realizations of regular maps.

9.4   On the Steinitz problem for tori

This section is devoted to the genus one case of Problem 9.1, the so-called *Steinitz problem for tori*. We are going to look at this problem, hoping that methods from oriented matroid theory cast new light on this longstanding open problem. Our methods are so general that even after a solution of the Steinitz problem for tori we have open cases for higher genus.

We next introduce some notation.

Definition 9.3 *A two-dimensional closed manifold is a topological space such that each point has a neighborhood that is homeomorphic to the interior of a two-dimensional Euclidean ball. We assume our 2-manifolds to consist of only one connected component.*

A non-mathematician should think in the two-dimensional case of a rubber membrane surface without boundary. Handles and holes are allowed. We study this object without distances on the surface and even without seeing it as some usual geometric object in our space. However, the connectivity plays a substantial role. A manifold can of course carry additional mathematical structure such as metrical properties if we investigate polyhedral structures. The manifold can even have smoothness properties that are important in differential geometry.

Topological 2-manifolds are well understood in mathematics. Topology tells us that they are either oriented 2-manifolds in which case they are 2-spheres with a finite number $g = 0, 1, 2, \ldots$, its genus, of handles attached to the 2-sphere, or they are unoriented 2-manifolds in which case they are 2-spheres with a finite number $\hat{g} = 1, 2, \ldots$, its cross cap number, of Möbius strips inserted in circular holes on the 2-sphere, see for example Fomenko, 1994.

Starting with an abstract point set $E = \{1, 2, \ldots, n\}$, we define a set of abstract triangles T as a family of three element subsets of E. We call a two element subset of such an abstract triangle $t \in T$ an *(abstract) edge*, and we require it to occur precisely twice as a subset of an abstract triangle. Consider all triangles that are incident with a fixed abstract point. We require that those edges not containing the fixed abstract point, form a cyclic sequence, that is, we can write these edges in the form $\{a, b\}, \{b, c\}, \ldots \{., a\}$. We require that we have pairwise different elements $a, b, \ldots \in E$. We assume that for any two points, there exists a set of edges connecting these points.

Definition 9.4 *We call such a list of abstract triangles an (abstract) triangulated (connected) closed 2-manifold.*

If we assign to each abstract point a corresponding point in Euclidean 3-space, we can consider edges and triangles as corresponding convex hulls of these points in 3-space. In the orientable case of closed 2-manifolds, we use the following concept.

Definition 9.5 *A list of triangles of an oriented (abstract) triangulated closed 2-manifold forms a polyhedral triangulated (connected) 2-manifold embedded in*

Euclidean 3-space, or an embedding, if there is a mapping of abstract points to corresponding points in Euclidean 3-space such that any pair of triangles has no common inner point.

It is known that unoriented closed 2-manifolds cannot be embedded in Euclidean 3-space. However, in this unoriented case the following concept of an immersion can occur.

Definition 9.6 *A list of triangles of an unoriented (abstract) triangulated closed 2-manifold forms a polyhedral triangulated (connected) 2-manifold, immersed in Euclidean 3-space, or an immersion, if there is a mapping of abstract points to corresponding points in Euclidean 3-space such that any pair of triangles with a common vertex has no common inner point.*

Later, in Section 9.5.3, we will look at the problem of finding immersions for non-orientable 2-manifolds as well. However, this section is devoted to the *torus* case, to an oriented closed 2-manifold of genus one. With the above concepts we formulate the *Steinitz problem for tori* once more.

Problem 9.7 *Can we find for each abstract oriented triangulated closed 2-manifold of genus one a corresponding embedding in 3-space?*

We start with an example of a torus with nine vertices. In Figure 9.14 we find its combinatorial description.

We start with a vertex with the highest valence. Point 4 is incident with eight edges. So we start with a regular octagon and consider it as the boundary of a cone with apex 4. Point 4 should lie far below the center of the regular octagon. Now

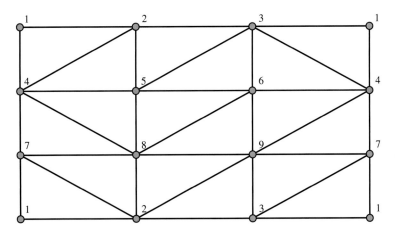

Figure 9.14   A triangulated torus with nine vertices

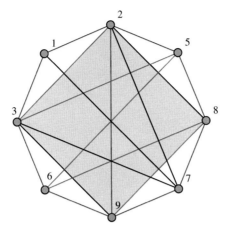

Figure 9.15 A triangulated torus with nine vertices, realization

we insert triangles. We see that triangles [2,3,9] and [2,8,9] form a quadruple that we can consider to be nearly planar. If we lift vertices 1 and 7 and we put points 5 and 6 below the nearly quadrangle we started with, we have found a realization. Very often, similar to this example, a particular realization can be found easily and it can also be described easily.

However, can we find general methods for this and related problems? Given an abstract triangulated 2-manifold and deciding whether there exists a corresponding embedding or a corresponding immersion as a polyhedral triangulated 2-manifold, is indeed a difficult problem. Our general method consists of splitting the problem into two subproblems each of which is again very difficult to solve in the general case. The first subproblem is to find an admissible oriented matroid. The second subproblem is to find a corresponding realization of the oriented matroid or a proof that such a realization cannot exist. However, since both these subproblems are very difficult in the general case, heuristic methods play a major role in this context.

Attempts of Grünbaum, Duke, Mani-Levitska, Lavrenchenko and others to solve the Steinitz problem for tori have led to a list of 21 unshrinkable tori which are known to be realizable. By *shrinking an edge* of a torus, one understands the deletion of two adjacent triangles in the map of the torus such that the resulting map, with two triangles less, is again a torus.

9.4.1 A triangulated torus

We consider another triangulated torus as an example. We have a metrical symmetry of order 6. Vertices are those of a regular octahedron together with three vertices of a regular triangle, compare Figure 9.16. The triangles of the torus are

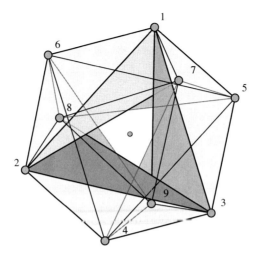

Figure 9.16   Projection of a triangulated torus

those of the regular octahedra except two triangles of opposite sides. A plane
in-between these deleted triangles and parallel to them contains the additional
triangle that defines the remaining three vertices. The midpoints of these three
missing triangles lie on a line. A line from the midpoint of the octahedron through
one of the additional vertices meets an edge of the octahedron orthogonal and in
its midpoint.

The list of all triangles is the following

1	2	6		2	3	4		1	3	5		1	5	6	2 4 6	3 4 5
1	3	9		1	2	7		2	3	8
2	7	8		1	7	9		3	8	9
6	8	9		4	7	9		5	7	8
5	6	8		4	6	9		4	5	7

This was again an example where we have a realization of the triangulated
2-manifold. The inverse problem is the difficult one.

We discuss as an exercise for a Haskell program the problem of deciding
whether the list of triangles is that of a closed 2-manifold. We first test the pseudo-
manifold property, that is, every edge lies is precisely two triangles. In addition
we require the following. For all triangles that are incident with a fixed vertex,
we require that their edges, that do not have this fixed vertex as an endpoint,
form a finite number of cycles without repeated vertices in a cycle. If we have
precisely one cycle for each vertex, we have a closed 2-manifold, otherwise, we
call it a pseudo-manifold.

The following Haskell program should be easy to understand.

Triangle list of a Klein bottle

```haskell
kleinBottle::[[Int]]
kleinBottle
 =[[1,2,7],[1,2,8],[1,3,4],[1,3,5],[1,4,7],[1,5,8],
   [2,3,6],[2,3,9],[2,6,7],[2,8,9],[3,4,6],[3,5,9],
   [4,5,7],[4,5,8],[4,6,8],[5,7,9],[6,7,9],[6,8,9]]
pseudoManifoldTest::[[Int]]->Bool
pseudoManifoldTest triangleList
   =   edgeTest triangleList
     &&(and [twoVertices (concat star)
          | star<-starList])
   where
   edgeTest::[[Int]]->Bool
   edgeTest triangleList
     = twoEdges (concat (map triangle2Edges
             triangleList))
     where
     twoEdges::[[Int]]->Bool
     twoEdges [] = True
     twoEdges list@(x:xs)
        | x 'elem' xs && x 'notElem' (xs\\[x])
          = twoEdges (xs\\[x])
        | otherwise   = False
   triangle2Edges::[Int]->[[Int]]   -- a < b < c
   triangle2Edges [a,b,c] = [[a,b],[b,c],[a,c]]


   starList = [[(triangle\\[vertex])
                  |triangle <- triangleList,
                   vertex 'elem' triangle ]
                |vertex<- nub (concat triangleList) ]
   twoVertices::[Int]->Bool
   twoVertices      [] = True
   twoVertices list@(x:xs)
      | x 'elem' xs && x 'notElem' (xs\\[x])
         = twoVertices (xs\\[x])
      | otherwise   = False
> pseudoManifoldTest kleinBottle
  True
```

The following functions generate all cycles of all links. If there is precisely one cycle, we obtain a 2-manifold instead of just a pseudo-manifold.

```
linkGen::Int->[[Int]]->Int->[Int]->[[Int]]->(Int,[[Int]])
linkGen i cyclesSoFar _  cycleSoFar@c  []
      = (i,cyclesSoFar++[c])
linkGen i cyclesSoFar@cSF startVertex@v
        cycleSoFar@c
        remainingEdgeList@(e:es)
  |last c == v = linkGen i (cSF++[c]) (head e) e es
  |last c == head e
    = linkGen i cSF v (cycleSoFar++[last e]) es
  |last c == last e
    = linkGen i cSF v (cycleSoFar++[head e]) es
  |otherwise
    = linkGen i cSF v  cycleSoFar     (es++[e])

allLinks::[[Int]]->[(Int,[[Int]])]
allLinks triangleList
   = [ let star = [(triangle\\[vertex])
                  |triangle <- triangleList,
                   vertex 'elem' triangle ]
       in linkGen vertex [](head (head star))
          (head star)(tail star)
          vertex<- nub (concat triangleList) ]

--sort (allLinks kleinBottle)
--   [(1,[[2,7,4,3,5,8,2]]),
--    (2,[[1,7,6,3,9,8,1]]),
--    (3,[[1,4,6,2,9,5,1]]),
--    (4,[[1,3,6,8,5,7,1]]),
--    (5,[[1,3,9,7,4,8,1]]),
--    (6,[[2,3,4,8,9,7,2]]),
--    (7,[[1,2,6,9,5,4,1]]),
--    (8,[[1,2,9,6,4,5,1]]),
--    (9,[[2,3,5,7,6,8,2]])]

manifoldTest::[[Int]]->Bool
manifoldTest triangleList
   = pseudoManifoldTest triangleList
      && linkNumberTest triangleList
```

```
linkNumberTest::[[Int]]->Bool
linkNumberTest triangleList
   = and[length(snd((allLinks triangleList)!!j))==1
        | j<-[0..(m-1)]]
   where
   m = length (allLinks triangleList)

> manifoldTest kleinBottle
  True
```

9.4.2 The unshrinkable tori

Grünbaum and Duke (unpublished) and Lavrenchenko, 1987, independently, have determined the unshrinkable triangulated tori depicted in Figure 9.17, compare also a recent survey article by Negami, 2004.

Corresponding arrows have to be identified in all these cases in order to obtain a torus. All the remaining triangulated tori can be obtained by an inverse edge shrinking process called vertex-splitting.

The 21 unshrinkable tori can all be found as subcomplexes in the 2-skeleton of 4-polytopes. Schlegel diagrams of the 4-polytopes tell us that these tori are realizable in 3-space without self-intersections. We have according to Euler's formula $f_0 + f_1 + f_2 = 2 - 2g$, $g = 1$, and by counting incidences of edges and triangles $3f_2 = 2f_1$. For (f_0, f_1, f_2) we must have $\binom{f_0}{2} \geq f_1$. This is violated for $f_0 < 7$. For $f_0 = 7$ there is precisely one solution. The f-vectors of the unshrinkable tori are $(f_0, f_1, f_2) = (k, 3k, 2k)$ with $k \geq 7$.

Let l_i be the number of vertices in the link of vertex i, $l_1, l_2, \ldots, l_k$. We have $\sum_{i=1}^{i=k} l_i = 6k$. The average number of vertices in a link is 6. We can continue to collect properties of a minimal triangulated non-realizable torus. We see, for example, that a minimal non-realizable torus contains no link with three vertices. Otherwise, we can find an even smaller torus in which the link has been filled with a triangle.

If a torus lies as a subcomplex in the 2-skeleton of a 4-polytope, we find a realization via a Schlegel diagram of the polytope. If we look for a candidate which cannot be embedded, we can look at examples of non-polytopal 3-spheres.

In the discussion of Altshuler's sphere No. 963, we have seen how a Haskell program can be used to find tori within the 2-skeleton of a simplicial 4-polytope. If we find such a corresponding 4-polytope, we have found a realization of the torus. Any Schlegel diagram would lead to such a realization.

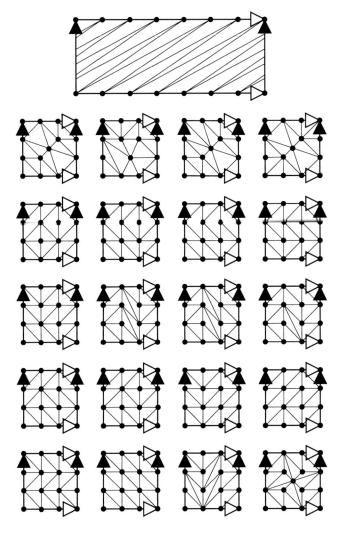

Figure 9.17   The unshrinkable triangulated tori of Grünbaum and Duke

9.4.3   *Triangulated tori with 11 vertices are realizable*

The idea of embedding the torus into the 2-skeleton of a 3-sphere (hopefully into that of a polytopal 3-sphere) was pursued in an investigation by Bokowski and Fendrich. As a result we have the following theorem.

Theorem 9.8 *[Bokowski and Fendrich] All triangulated tori with up to 11 vertices can be realized.*

The proof starts with an investigation by Bokowski and Strempel showing that all unshrinkable tori can be embedded in the 2-skeleton of a convex 4-polytope.

This result was also earlier obtained by Grünbaum. Remaining arguments use all inverse edge shrinkings and investigations as to whether the torus lies in a 2-skeleton of a 4-polytope. Finally, a small remaining set of tori was realized directly.

If this local argument can be improved such that one remains always within the class of polytopal 3-spheres, we have via any Schlegel-diagram of the 4-polytope, a realization of the torus without self-intersections.

In contrast to the above result, we know a simplicial oriented 2-manifold, compare Altshuler, Bokowski, and Schuchert, 1996, which cannot be embedded in the 2-skeleton of any oriented matroid polytope. We come to this example later.

9.5 Spatial polyhedra

9.5.1 Polyhedra with small genus

The Steinitz problem for tori is at the moment open and challenging. Can we solve the corresponding higher genus cases? This is indeed the case for genus $g \geq 6$. We will explain the corresponding result later in Section 9.7.

So far, all attempts to find a non-realizable triangulated orientable polyhedron of genus g with $0 < g \leq 5$ have failed. We can try to find for each genus realizable examples with the smallest number of vertices. In addition, we can ask for the most symmetrical realizations. The list of Brehm in Table 9.1 shows an overview.

Triangulated closed polyhedra with up to ten vertices have been investigated by Bokowski and Lutz. These 2-manifolds contain orientable ones up to genus 3. In all these cases, the C++ extension program for rank 4 oriented matroids has been used to find compatible oriented matroids. A further investigation for realizability by Lutz and coauthors has led to realizable ones in all cases.

Theorem 9.9 *[Bokowski and Lutz] All triangulated orientable closed 2-manifolds with up to ten vertices have admissible oriented matroids, that is, a realization of these oriented matroids would lead to realizable polyhedra without self-intersections in 3-space.*

9.5.2 Pinched neighborly two-manifolds

The investigation in Altshuler, Bokowski, and Schuchert, 1994 considers 2-dimensional pseudo-manifolds. These objects have 2-manifold properties up to a finite number of points where the star property has been replaced by a finite number of stars with a common vertex. If we replace such a finite number of

Table 9.1 *List of triangulated closed orientable polyhedra of genus*
g with minimal number of vertices f_0

g	f_0	existence	maximal symmetry
0	4	trivial	S_4
1	7	Császár, 1949	Z_2
2	10	Brehm, 1981	$Z_2 \times Z_2$
3	10	Brehm, 1987	$Z_2 \times Z_2$
4	11	Bokowski and Brehm, 1989	Z_2
5	12	open	
≥ 6	12	Bokowski and Guedes de Oliveira, 2000; non-realizable ones exist	

stars with a common vertex by a set of stars by using as many vertices as we have closed cycles, we obtain a 2-manifold.

We can embed a complete graph with ten vertices on a 2-sphere without self-intersections if we allow some pointwise identifications of points on the sphere. The resulting pinched sphere forms a pseudo-manifold. There are four combinatorially different examples. The following triangle list is one of them.

```
pinchedSphere::[[Int]]
pinchedSphere
 = [[1,2, 3],[1,2, 4],[1,3, 8],[1,4,10],[1,5, 8],
    [1,5, 9],[1,6, 7],[1,6, 9],[1,7,10],[2,3, 9],
    [2,4,10],[2,5, 6],[2,5,10],[2,6, 8],[2,7, 8],
    [2,7, 9],[3,4, 5],[3,4, 9],[3,5, 7],[3,6, 7],
    [3,6,10],[3,8,10],[4,5, 6],[4,6, 8],[4,7, 8],
    [4,7, 9],[5,7,10],[5,8, 9],[6,9,10],[8,9,10]]
```

The function `pseudoManifoldTest` from Page 233 applied to `pinchedSphere` returns `True`. The function `manifoldTest` from Page 234 applied to `pinchedSphere` returns `False`.

Testing the pseudomanifold or the manifold property with our Haskell functions does not check the connectivity. Figure 9.18 shows the combinatorial properties of our example.

Can we find a model with flat triangles and without self-intersections? For solutions we refer the reader to Altshuler, Bokowski and Schuchert, 1994. We show one example from this article in Figure 9.19 as an exploded view of the polyhedron.

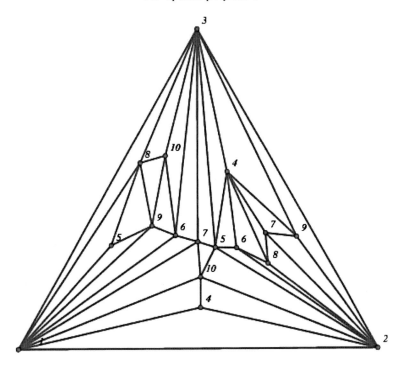

Figure 9.18   Triangulation of a pinched sphere with ten vertices

9.5.3   Immersions of 2-manifolds

For a flat *immersion* of a triangular combinatorial 2-manifold, we require that any pair of triangles that is incident with a common point has no common inner point. In other words, each star of a vertex has no self-intersection. The oriented matroid condition for this is easily formulated. The list of forbidden circuits given by pairs of a triangle and an edge is smaller compared with the stronger condition that we encounter in embedding problems for triangular oriented 2-manifolds. The generation of oriented matroids under the restriction of these forbidden circuits would lead to admissible oriented matroids if they exist.

However, clever heuristic methods, if successful, do the job of finding realizations as well. Such successful heuristic findings are essential results in the articles of D. Cervone. He, 1994, deals for example with immersions of the Klein bottle and solves a corresponding question completely.

Theorem 9.10   *[Cervone] Any simplexwise-linear simplicial immersion of the Klein bottle into* $\mathbb{R}^3$ *has at least nine vertices and there exist examples in each immersion class with exacly nine.*

Figure 9.20 shows one of his examples.

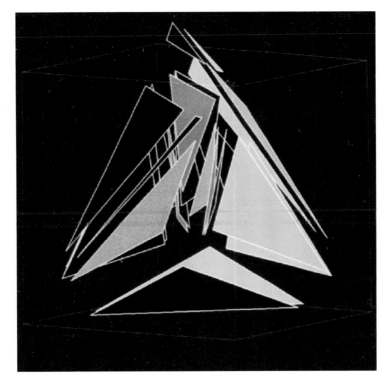

Figure 9.19 A pinched 2-manifold, exploded view

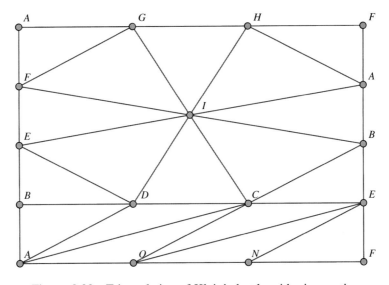

Figure 9.20 Triangulation of Klein's bottle with nine vertices

We list coordinates of Cervone for a flat immersion of this triangulation in Figure 9.20 of the Klein bottle.

$$
\begin{array}{ll}
A & (1 - \sqrt{3}, 0, 0.5) \\
B & (\sqrt{3} - 1, 0, 0.5) \\
C & (-1, 1, -0.25) \\
D & (-1, -1, 0.25) \\
E & (\sqrt{3} - 1, 0, -0.25) \, . \\
F & (1 - \sqrt{3}, 0, -0.25) \\
G & (1, 1, 0.25) \\
H & (1, -1, -0.25) \\
I & (0, 0, 0)
\end{array}
$$

Let us summarize results of Cervone in the case of immersions of 2-manifolds. A combinatorial part deals with the generation of all abstract triangulated 2-manifolds with a small number of vertices, for example, for the Klein bottle. A later argument shows that none of those with eight vertices can be immersed in 3-space. On the other hand he shows examples of 9-vertex immersions.

In Cervone, 1996, he considers the special case of when the polyhedral manifolds in question are so-called *tight* ones. This notion, introduced by N. H. Kuiper, who used this concept in differential geometry, concerns a subclass of manifolds with a certain natural property. Tightness means that the intersection of the manifold with any half-space remains connected. We can say that the manifold is *as convexly as possible*. We find tight simplicial realizations in Cervone's articles that are not at all easy to find, see Cervone 1997, 2000, 2001.

In conclusion, we remark that such realizations are now seen within the framework of general methods obtainable from oriented matroid techniques.

9.5.4 Polyhedral realizations of regular maps

In the transition from Platonic solids to regular maps, we replace the flag-transitive symmetry groups, with rigid motions as elements, by flag-transitive symmetry groups on a purely combinatorial level.

In a recent survey article about maps, Vince, 2004, we read that maps are probably the oldest topic in graph theory. Regular maps and their polyhedral realizations are a natural subclass within the study of graphs on surfaces. We mention some results. We can see these results within the framework of techniques that we have learned from the theory of oriented matroids.

For a survey on this subject, we also refer the reader to Coxeter and Moser, 1980 and Wilson, 1976. For a more recent profound investigation of vertex transitive maps, see Lutz, 1999.

Figure 9.21   Klein bottle, pottery model

A series of papers deal with the question of realizing these regular maps as poly-hedra. We mention for example Wills, 1987, Bokowski and Wills, 1988. Among these realizations self-intersections are sometimes allowed as in the famous Kepler–Poisot bodies. We have results of Schulte, Wills, Bokowski, Brehm, and others in this area.

For a first spatial polyhedral representation of Dyck's regular map $\{3, 8\}_6$, see Bokowski, 1989. This inspired Brehm and he found a maximal symmetric polyhedral realization, see Brehm, 1987.

We come back to regular maps when we study curve arrangements on 2-manifolds.

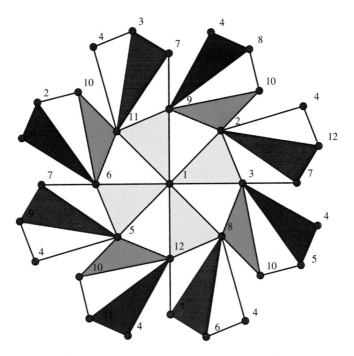

Figure 9.22 Dyck's regular map $\{3, 8\}_6$

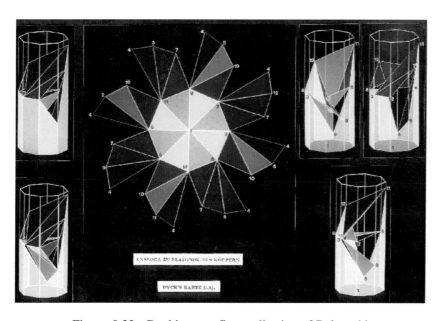

Figure 9.23 Dyck's map, first realization of Bokowski

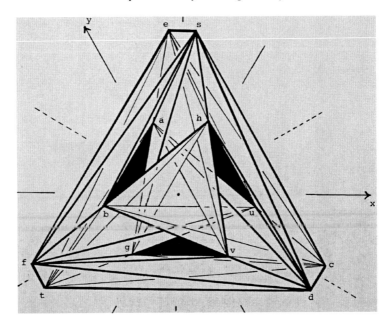

Figure 9.24 Brehm's symmetric realization of Dyck's map

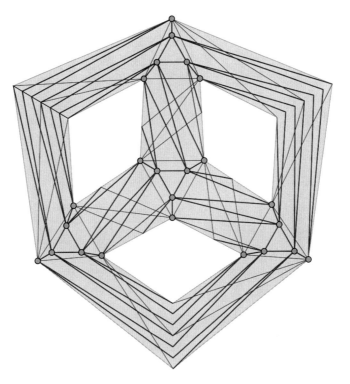

Figure 9.25 Klein's regular map $\{3, 7\}_8$ on a genus 3 surface

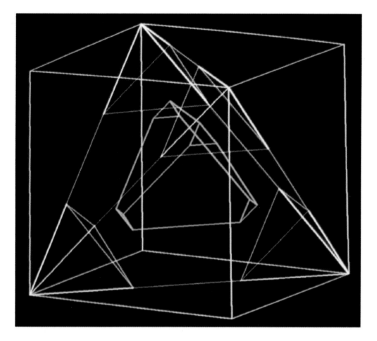

Figure 9.26   Felix Klein's regular map, vertex set of polyhedral realization without self-intersections of Schulte and Wills

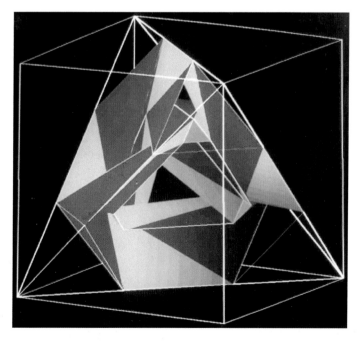

Figure 9.27   Felix Klein's regular map, polyhedral realization of Schulte and Wills, 1985

9.6   Non-realizable polyhedra

Sometimes the search for flat embeddings of 2-manifolds without self-intersections in Euclidean 3-space has no solution.

9.6.1   *A non-realizable torus with seven vertices*

When we call the Möbius torus *seven-vertex torus*, this might lead to confusion. There exists another seven-vertex torus when we allow more than one edge connecting two vertices, see Figures 9.28 and 9.29. In both cases we have an equal number of triangles, edges, and vertices.

Of course, the triangulated torus of the pottery model in Figure 9.28 cannot be realized with flat triangles as a triangulated torus.

9.6.2   *Simutis's non-realizable torus*

When we consider 2-manifolds that are not triangulated, we have non-realizable examples.

The torus in Figure 9.30 cannot be realized with convex faces, see Simutis, 1977. In Bokowski and Sturmfels, 1989b, p. 72 ff., we find a proof that uses the

Figure 9.28   Another seven-vertex torus, pottery model

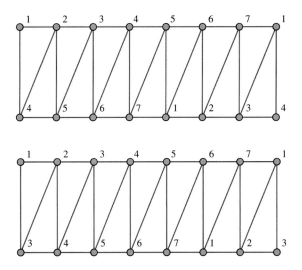

Figure 9.29   Two different triangulated tori with seven vertices, Möbius torus (above) and pottery model (below)

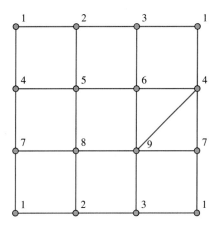

Figure 9.30   Simutis's example of a non-realizable torus with flat faces

final polynomial method of oriented matroid theory, showing that a realization cannot exist even if we allow non-convex faces. Although, the polynomial covers half a page, the argument is simple.

We did not allow the two triangles that are incident with edge 4, 9 to lie in the same plane. Otherwise, the torus is clearly realizable.

9.6.3   Brehm's Möbius strip

There exists a triangulated Möbius strip that cannot be embedded in 3-space without self-intersections, see Brehm, 1983. Figure 9.31 shows the triangulation

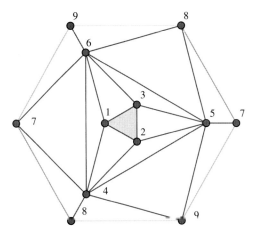

Figure 9.31   Brehm's triangulated Möbius strip, combinatories

Figure 9.32   Brehm's triangulated Möbius strip, pottery model

as a part of the projective plane, opposite edges have to be identified. The inner triangle 1,2,3 is missing. The representation in Figure 9.31 does not indicate that there might be problems if we look for an intersection free realization with flat triangles. However, the pottery model of a Möbius strip with the corresponding triangulation with arcs in Figure 9.32 provides a different impression.

In an investigation of Guedes de Oliveira and Carvalho, 2002, they have obtained, along Brehm's arguments but on a purely oriented matroid level, a more general result.

9.7 Non-realizable triangulated orientable surfaces

In the plane or on the 2-sphere four different points can be connected pairwise with curves such that no pair of curves intersects. We have drawn the complete graph with 12 vertices in Figure 9.33 to see that in this case there are many crossings in the plane.

We have seen that we can connect on the torus seven different points pairwise with curves such that no pair of curves intersects. We have discussed that example, the Möbius torus, in the previous section. We have also mentioned Heawood's inequality for the chromatic number $\chi(S_g)$ of an orientable closed surface S_g of genus g on Page 223. Ringel and Youngs have proved the equality sign

$$\chi(g) = \lfloor \frac{1}{2}(7 + \sqrt{48g + 1}) \rfloor \text{ for } g \geq 1$$

This is the celebrated map color theorem of Ringel and Youngs, see Ringel, 1974. The four-color problem, proved in 1976, is the case $g = 0$. See also Robertson, Sanders, Seymour, and Thomas, 1997.

When we use a map that needs the maximum number of colours, we have with its dual map a triangular complete graph embedding. In other words, we can connect 12 points on a genus 6 surface by arcs without crossings. However, a solution is not easy to find by hand. Can you connect 12 points pairwise without

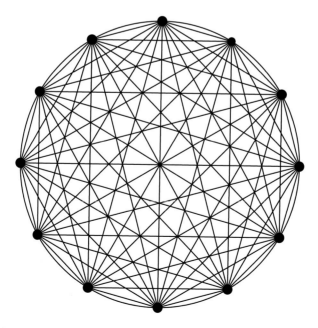

Figure 9.33   Complete graph with 12 vertices in the plane

crossings on a surface of genus 6, that is, on a surface that has no boundary and that containes no Möbius strip?

We find a first triangular complete graph embedding with 12 vertices on a closed orientable surface of genus $g = 6$ in Heffter, 1891. The following scheme of Heffter can be interpreted as follows.

Heffter's complete graph embedding

```
 1.   2  3   4   5   6   7   8   9  10  11  12
 2.   3  1  12   4   9   8   5  11   7  10   6
 3.   4  1   2   6   5   8  10   7  12  11   9
 4.   5  1   3   9   2  12   8   7  11   6  10
 5.   6  1   4  10  12   7   9  11   2   8   3
 6.   7  1   5   3   2  10   4  11   8  12   9
 7.   8  1   6   9   5  12   3  10   2  11   4
 8.   9  1   7   4  12   6  11  10   3   5   2
 9.  10  1   8   2   4   3  11   5   7   6  12
10.  11  1   9  12   5   4   6   2   7   3   8
11.  12  1  10   8   6   4   7   2   5   9   3
12.   2  1  11   3   7   5  10   9   6   8   4
```

We connect the vertex in the first column in the following circular sequence – orientation of this sequence should be the same for all vertices – with the remaining vertices by edges. It turns out that these sequences are compatible. As a result we obtain a complete graph embedding with 12 vertices on a surface of genus 6.

Again, imagine that the surface of genus 6 is given as a model like the pottery model in Figure 9.34. Even when you know Heffter's scheme, drawing the 44 edges as curves on the surface is not an easy task.

We are going to provide a solution for this drawing problem later. However, we use a different map with 44 triangles, 66 edges and 12 vertices.

Finding all complete graph embeddings

Altshuler has found altogether 59 different maps of complete graph embeddings on a surface of genus 6. Bokowski has shown this list to be complete, see Altshuler, Bokowski, and Schuchert, 1996.

Altshuler used local argument to construct new maps from known maps. So he was not sure whether he had found all these maps.

Generating the start stars

```
sts::[(Int,[[Int]])]
sts=[(i,(map(\k->[mod((head k)+i)11]))
      ([[10]]++[[j]|j<-[2..9]]++[[1]]))|i<-[0..10]]
```

Figure 9.34   Can you connect 12 points pairwise by arcs without crossings?
Pottery model of a genus 6 surface

We provide the essential global argument of Bokowski for finding all 59 different maps in the following Haskell code. We sketch the underlying idea. We look at all stars of the map with vertices $0, 1, 2, \ldots, 11$. For the star with vertex 11 we can fix without loss of generality the labeling of the triangles. This implies that we know two adjacent triangles for all the other stars. We represent one of the growing eleven stars by a pair of its midpoint and a growing (so far undetermined) sequence of points along its link. The left bound and the right bound of this link is fixed from the beginning because of the two determined triangles. The inner points can be glued according to those triangles that we insert step by step during the algorithm. The function `sts` defines the start stars for the 11 growing stars.

Inserting a triangle

```
itr::[Int]->(Int,[[Int]])->(Int,[[Int]])
itr [a,b,c](j,gs)|j'notElem'[a,b,c]=(j,gs)
                 |j==a=ins[b,c](j,gs)
                 |j==b=ins[c,a](j,gs)
                 |j==c=ins[a,b](j,gs)
```

The idea is now to find possible triangles that are consistent with the growing star structure. When we look at the left bound of one star, there are possible points that can be glued to this bound. Later on there are less options for gluing these points along a link because we insert all triangles immediately in all the other stars. The function `itr` inserts such triangles in the growing stars.

When we have for each star a fixed sequence as its link, we have a map.

Extending the star structure

```
e::[[(Int,[[Int]])]]->[[(Int,[[Int]])]]
e [] = []
e (s:r)
 |[]'elem'(map snd)s = e r
 |(map length)((map snd)s)
   ==[1,1,1,1,1,1,1,1,1,1,1] = [s]++(e r)
 |length gst==2=[(map(itr[m,ly,head(lagst)])))s]++(e r)
 |otherwise
   = [(map(itr[m,ly,head(gst!!i)])))s|i<-[1..lgst]]++(e r)
 where
 m  = minimum[i|i<-[0..10], length(snd(s!!i))/=1]
 gst= snd (s!!m); lgst=(length gst)-2; lagst=last gst
 y  = head gst; ly = last y; ys = tail gst
```

Inserting a line segment in a star

```
ins::[Int]->(Int,[[Int]])->(Int,[[Int]])
ins [a,b] (j,gs)
 |length gs==1 = (j,[])
 |length gs==2&&a==last(head gs)&&b==head(last gs)
 =(j,[new])
 |length gs==2 = (j,[])
 |a==last(head gs)&&b'elem'(map head)(tail(init gs))
 =(j,[new]++all)
 |a'elem'(map last)(tail(init gs))&&b==head(last gs)
 =(j,all++[new])
 |  a'elem'(map last)(tail(init gs))
  &&b'elem'(map head)(tail(init gs))
   =(j,(fst(splitAt 1 all))++[new]++(snd(splitAt 1 all)))
 |otherwise=(j,[])
 where all=[el|el<-gs,el/=beg,el/=end]
       beg=head[el|el<-init gs,last el==a]
       end=head[el|el<-tail gs,head el==b]
       new=beg++end
```

We find a first map via `take 1 ((iterate e [sts])!!34`. We apply the function `e` 34 times since the map has 44 triangles altogether, 11 triangles are known at the beginning and we have to be sure that we have checked the final version. The first map of our Haskell program is the following.

```
[[(  0,[[10,2,4,3, 5,7, 9, 6,8, 1]]),
  ( 1,[[ 0,8,3,9, 5,4, 6,10,7, 2]]),
  ( 2,[[ 1,7,5,9, 8,4, 0,10,6, 3]]),
  ( 3,[[ 2,6,9,1, 8,7,10, 5,0, 4]]),
  ( 4,[[ 3,0,2,8,10,9, 7, 6,1, 5]]),
  ( 5,[[ 4,1,9,2, 7,0, 3,10,8, 6]]),
  ( 6,[[ 5,8,0,9, 3,2,10, 1,4, 7]]),
  ( 7,[[ 6,4,9,0, 5,2, 1,10,3, 8]]),
  ( 8,[[ 7,3,1,0, 6,5,10, 4,2, 9]]),
  ( 9,[[ 8,2,5,1, 3,6, 0, 7,4,10]]),
  (10,[[ 9,4,8,5, 3,7, 1, 6,2, 0]])]]
```

From all maps of our Haskell program, we have finally to discard repeated solutions that occur because of many symmetric maps.

For the map that we have used in Bokowski and Guedes de Oliveira, 2000, we have depicted all triangles in Figure 9.35. There is a combinatorial symmetry of order 6.

The solution of the graph embedding on the surface can be seen in Figure 9.36. However, the surface is no longer easily seen. The Goose Neck model of the author received its name because of its goose neck material. The model was presented for the first time in Bydgoszcz, Poland in 1998.

The model shows 12 points in space. Any pair of points is connected by a curve, called an edge. The 12 points and 66 edges form a complete graph with 12 vertices. The model has a 3-fold symmetry with a vertical axis of symmetry. An additional symmetry allows us to put the model upside down without changing its shape. From each point 11 edges start. At the beginning they define a plane near the point from which they start. If we follow two neighboring edges, we find two adjacent positions at the other points from which the same additional edge starts. Three such edges form a frame which can be considered to contain a surface, a topological triangle. The surface consisting of all these topological triangles has 44 triangles altogether. Each edge is adjacent to two triangles. At each vertex the triangles form a cycle. We have a triangulated surface without boundary. On this surface, the 66 edges lie without self-intersections.

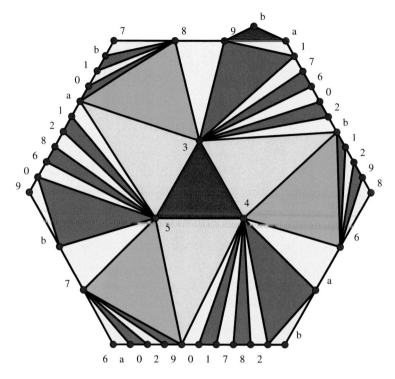

Figure 9.35 Forty-four triangles forming an embedded complete graph on a genus 6 surface

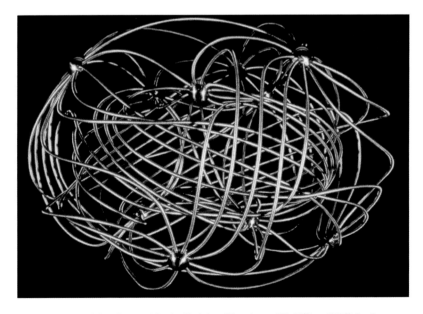

Figure 9.36 Goose Neck, Polska 98, photo N. Hähn, Wiësbaden

Figure 9.37   Top view of the gooseneck model of Figure 9.36

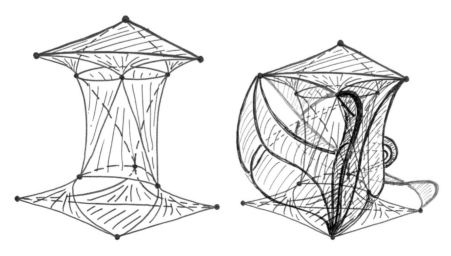

Figure 9.38   Some parts of the surface

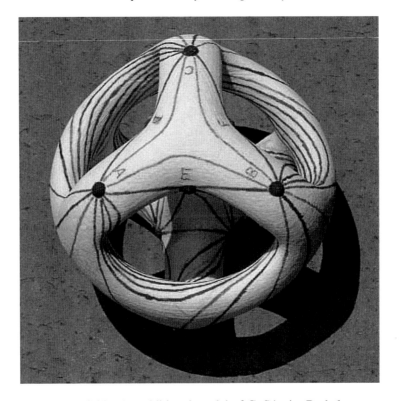

Figure 9.39   An additional model of C. Séquin, Berkeley

We have to imagine the remaining parts of the surface. Recall that there is a rotational symmetry of order 3 and the model does not change when we put it upside down. Some parts of the surface have been depicted in Figure 9.38 and its top view in Figure 9.37. For this example we have:

Theorem 9.11 *[Bokowski and Guedes de Oliveira] This triangulated closed orientable 2-manifold cannot be embedded in 3-space with flat triangles and without self-intersections.*

The result, see Bokowski and Guedes de Oliveira, 2000, was unknown for a long time. The result required a total amount of over ten years of CPU-time. It is not the result itself which is surprising, but the method, based on the theory of oriented matroids.

We finally show a polyhedral model with the complete graph with 12 vertices in Figure 9.40, a corresponding model of C. Séquin in Figure 9.39, and a computer graphic in Figure 9.41. The list of triangles in this case is not isomorphic to the one used for the goose neck model. There are altogether 59 different non-isomorphic ways to do this, see Altshuler, Bokowski, and Schuchert, 1996, and by cutting

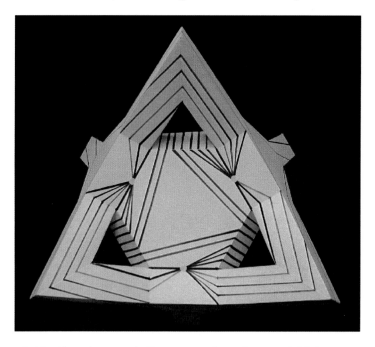

Figure 9.40   Complete graph K_{12} on a surface of genus 6, highest symmetry

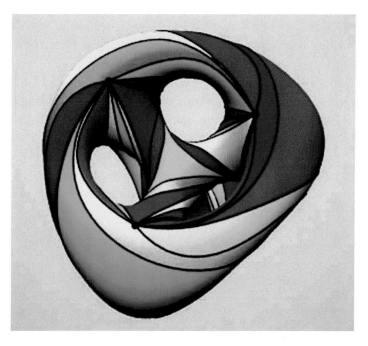

Figure 9.41   Complete graph K_{12} on a surface of genus 6, highest symmetry, computer graphics by C. Séquin and L. Xiao, Berkeley

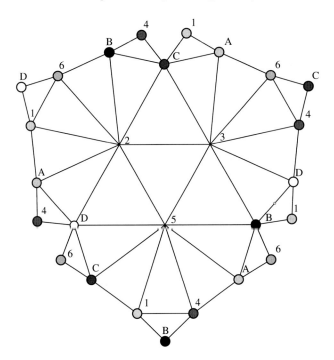

Figure 9.42 A realizable triangulated polyhedron of genus 3 with minimal number of vertices, see Bokowski and Brehm, 1985

a handle, twisting it around several times, and glueing it back again, we obtain even infinite many solutions.

Figures 9.39, 9.40, and 9.41 show a complete graph with 12 vertices on a surface of genus 6. This time we have chosen a map with the highest possible symmetry. It is unknown whether this map can be realized with flat triangles and without self-intersections. We invite the reader to find a first triangulated closed orientable 2-manifold with smaller genus than six that cannot be embedded in 3-space without self-intersections. Examples with minimal number of vertices for given genus were considered as candidates. However, the example of Figure 9.42 is realizable with planar triangles and without self-intersections.

10

Some oriented matroid applications

We have seen that oriented matroids are very useful in the theory of convex polytopes and polyhedral manifolds. However, already the variety of concepts that we have studied in the first two chapters have told us that applications of the theory of oriented matroids reach far beyond this scope. In this chapter we consider some additional applications in mathematical and non-mathematical fields.

10.1 Triangulations of point sets

We see in Figure 10.1 a triangulated convex polygon.

With the sharpened eye via oriented matroid concepts, we see that the set of these triangles can be interpreted as a subset of bases of the rank 3 oriented matroid. When we studied this observation more closely, we found that even an overview of all triangulations of this point configuration can be obtained on an abstract level. The metrical properties play no decisive role. Studying triangulations in this way is indeed an application of the theory of oriented matroids for which there is a lot to say, especially in higher dimensions.

There will be a forthcoming book on this subject by J. de Loera, J. Rambau, and F. Santos. We refer the reader to this exposition. We have already a monograph on this subject, see Santos, 2002.

10.2 Oriented matroid programming

The study of oriented matroids from a combinatorial perspective of linear programming is another topic that goes beyond what we have covered in this book. Bland's approach in his Ph.D. thesis, Bland, 1974, and Fukuda's work in his Ph.D. thesis, Fukuda, 1982, together with the work of Edmonds form the foundation in this area. Bland found a new and interesting pivot rule when he studied linear programming from a discrete perspective. This is worth emphasizing on a more general level.

259

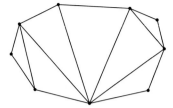

Figure 10.1   A triangulated convex hull in the plane

The oriented matroid stance leads to thinking in combinatorial terms. Forgetting metrical information allows a deeper analysis of the underlying problem. We refer the reader again to the book of Bachem and Kern for studying oriented matroids in the context of linear programming, see Bachem and Kern, 1992, However, see also Chapter 10 in Björner *et al.*, 1993, that is based on Fukuda's work.

10.3   Line configurations and algebraic surfaces

The book by Barthel, Hirzebruch, and Höfer, 1987, about line configurations and algebraic surfaces leads us beyond our scope. However, it shows that line configurations are a decisive tool in this context. So, the study of line configurations has unexpected applications.

10.4   Configurations of points and lines

In his book entitled *Geometrische Konfigurationen* from 1929, F. Levi wrote in his preface:

> *Überall, wo kombinatorische Methoden in der Geometrie fruchtbar werden, stößt man auf Konfigurationen, und ihr Studium wird so zur Einführung in die kombinatorische Geometrie.*

Within the study of configurations, oriented matroids can come into play as well. So, we are tempted to generalize Levi's remark by saying that the study of oriented matroids becomes an introduction to combinatorial geometry.

Configurations can be defined on a very general level. See Grünbaum, 2005, for a recent survey article about this topic. We consider here the special case of (connected symmetric) configurations. We drop the words connected and symmetric and define an (n_k)-configuration on an abstract level.

Definition 10.1 *A (connected symmetric) configuration (n_k) is a family of n (abstract) points and n (abstract) lines such that each of the points is incident with precisely k of the lines and each of the lines is incident with precisely k of* *the points. We consider only the connected case, that is, for any pair of points* (p_a, p_b) *there is an alternating sequence* $(p_a = p_1, l_1, p_2, l_2, \ldots, l_{k-1}, p_k = p_b)$ *of points and lines such that the points* p_i *and* p_{i+1} *are incident with the line* l_i, $i \in \{1, \ldots, k-1\}$. *The word symmetric compares these configurations with a more general concept.*

An (n_k)-configuration does not imply in general a realized point-line configuration in some Euclidean space or within some projective space.

10.4.1 Examples of (n_k)-configurations

Let us look at some examples of (n_k)-configurations. Points and lines of the Pappus configuration form a realizable (9_3)-configuration, compare with Page 108 and Page 267. There are altogether precisely three different (9_3)-configurations.

Figure 10.2 shows a realizable (27_3)-configuration.

In Grünbaum and Rigby, 1990, we find the realizable (21_4)-configuration that has been depicted in Figure 10.3.

At the end of that paper we find the description for a realizable configuration (28_4). Consider a perspective drawing of a cube in general position. Define 12 lines as the intersection of an arbitrary plane with one of the 12 planes that are spanned by four vertices of a cube. Define an additional 16 lines that are spanned by the edges and the main diagonals of the cube. As points we take the intersections of these 16 lines with the arbitrary plane, the vertices of the cube, the vanishing points of edges, and the midpoint of the cube. What we obtain has been depicted in Figure 10.4.

When we start with a simplex in an even dimensional Euclidean space $\mathbb{R}^{2k}$ with $2k+1$ vertices and we project it down onto a $k+1$-dimensional space we

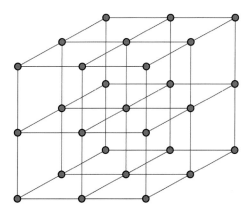

Figure 10.2 A (27_3)-configuration

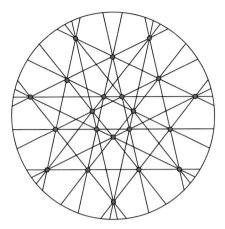

Figure 10.3   The real configuration (21_4) of Grünbaum that was studied in a more general context before by Klein, Burnside, Coxeter, and others

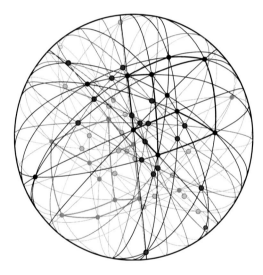

Figure 10.4   Configuration (28_4) in the sphere model of the projective plane

obtain $2k+1$ vertices in general position. Any $k+1$ points span a k-dimensional hyperplane and any k-points span a $(k-1)$-dimensional hyperline. When we use a plane in general position, these hyperplanes and hyperlines intersect this plane in lines and vertices that forms a $\left(\binom{2k+1}{k+1}_{k+1}\right)$-configuration. We obtain the well-known Desargue configuration in the special case $k=2$, compare with Page 270.

There is a long history of configurations, see Levi, 1929. In Bokowski and Sturmfels, 1989b, we see how oriented matroid techniques come into play. For (n_4)-configurations we find references when we look at Grünbaum, 2002, and the references cited there, see also `http://math.washington.`

`edu/˜grunbaum/index.html`. All realized examples appear to have been constructed in a particular way that would be difficult to find when we are given the combinatorial information alone.

10.4.2   Realizable (n_4)-configurations

A systematic collection of realizable (n_4)-configurations has been done by Grünbaum. We know realizable (n_4)-configurations for almost all n, however, we do not know realizable (n_4)-configurations for $n = 22, 23, 26, 29, 31, 32, 24, 37, 38,$ 43. Moreover, it has been conjectured that there is no realizable (n_4)-configuration for $15 \leq n \leq 20$. What can oriented matroid techniques do in this context? We have the following result.

Theorem 10.2 *[Bokowski and Schewe] There are no connected symmetrical* (15_4) *configurations and no connected symmetrical* (16_4) *configurations that can be realized in the plane. Moreover, there are not even pseudoline arrangements with this property.*

For $n = 15$ a contradiction follows from Euler's formula. For $n = 16$ the proof uses pseudoline arrangement properties, see Bokowski and Schewe, 2005.

10.4.3   An interesting (40_4) configuration

The problem of finding realizable (n_k)-configurations is not restricted to the rank 3 case. A particular (40_4)-configuration that has a high combinatorial symmetry played a role in the studies of H. Van Maldeghem (private communication). We write his combinatorial description of a (40_4)-configuration immediately as a Haskell code. We define points as 4-tuples with coordinates over GF_3. However, we identify negative tuples with the given one. Thus a reduction leads us to 40 points.

All 40 points

```
prepts::[[Int]]
prepts =[[a,b,c,d]|a<-s,b<-s,c<-s,d<-s,
          [a,b,c,d]/=[0,0,0,0]] where s = [-1,0,1]
reduce::[[Int]]->[[Int]]
reduce [] = []
reduce ([a,b,c,d]:xs)
  | [-a,-b,-c,-d]'elem' xs = reduce xs
  | otherwise              = [[a,b,c,d]]++reduce xs

pts = reduce prepts
```

We define a product between points. A pair of points lies by definition on a
line if and only if this product is zero modulo 3.

All 40 lines

```
prod::[Int]->[Int]->Int
prod [a1,b1,c1,d1] [a2,b2,c2,d2]
      = a1*b2 - b1*a2 + c1*d2 - d1*c2


on_ln::[Int]->[Int]->Bool
on_ln  p1 p2 =  (prod p1 p2 ==   0)
             ||(prod p1 p2 ==   3)
             ||(prod p1 p2 ==  -3)
lns::[[Int]]
lns = [[i,j,k,l]|i<-[0..36],    j <- [(i+1)..37],
                 k<-[(j+1)..38], l <- [(k+1)..39],
            on_ln(pts!!i)(pts!!j),on_ln(pts!!i)(pts!!k),
            on_ln(pts!!i)(pts!!l),on_ln(pts!!j)(pts!!k),
            on_ln(pts!!j)(pts!!l),on_ln(pts!!k)(pts!!l)]
```

Figure 10.5 depicts the corresponding incidence structure with points (corresponding to columns) and lines (corresponding to rows). We confirm that every
abstract point is incident with precisely four abstract lines and every abstract line
is incident with precisely four abstract points.

This configuration of points and lines has a realization over the field GF_3. We
can write down the matroid information as a list of unsorted hyperline configurations. However, deciding whether this matroid can be oriented is still a difficult
open problem. When we assume that we can find a realization in 3-space, we
can project the configuration of points and lines onto a plane in general position.
A study of the problem in the plane should be done first.

In the incidence structure of Figure 10.5 we can permute columns and rows. We
have done this in a particular way and have arrived at Figure 10.6. The lower right
corner structure reveals a (27_3)-subconfiguration. Realizing that configuration
first might be useful for deciding the original problem.

However, realizing an (n_3)-configuration is much easier than realizing an (n_4)-
configuration. In this particular case, a realization of the (27_3)-subconfiguration
was found by the author by using the dynamical drawing software *Cinderella*. In
similar problems this might lead to a conjecture of how difficult it might be to
realize a configuration.

We pick one example of those which can be described via the following
incidence table in Figure 10.7. When we can realize this configuration, we can

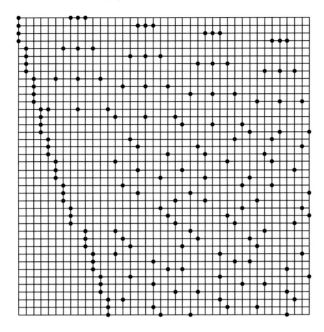

Figure 10.5 (40_4)-configuration of H. Van Maldeghem, can it be realized in 3-space or in the plane?

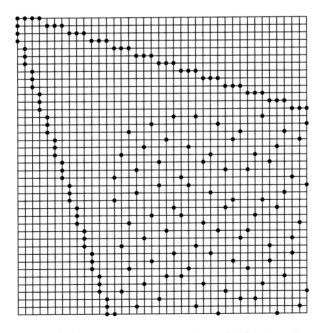

Figure 10.6 (40_4)-configuration revealing a (27_3)-subconfiguration

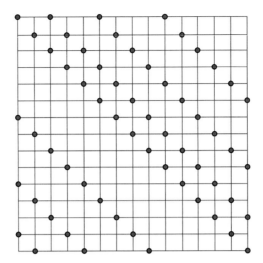

Figure 10.7 A (15_4)-configuration

find a rank 3 oriented matroid with 15 elements. The matroid is easy to describe. A point triple leads to a non-base precisely when all three points lie on a common line. Take a pair of bases $((i, j, k), (u, v, w))$ and delete one element of the first base. When the remaining two elements do not lie on a line, we can add any additional element from the second base to obtain again a base. Otherwise, two points lie on a common line. However, there exists an element in the second base that does not lie on this line. We see that the matroid property holds even for an arbitrary (n_k)-configuration.

10.5 Projective incidence theorems

A smallest non-stretchable pseudoline arrangement with nine elements was found by Levi, 1926. Levi applied Pappus's theorem, a well-known projective incidence theorem. Other projective incidence theorems, like Desargue's theorem, lead in a similar way to non-stretchable (non-realizable) oriented matroids. Proving that a pseudoline arrangement is not stretchable, is a challenging task. However, in many small example classes, final polynomials were found to confirm non-realizability of an oriented matroid. Whereas a projective incidence theorem can be used to show non-realizability of an oriented matroid, we can reverse the transition. We can use the final polynomial method to prove the non-realizability of an oriented matroid in the context of projective incidence theorems. We have obtained the following short proof of Pappus's theorem in this way.

10.5.1 Proof of Pappus's theorem

We assume that we have an ordered set of nine points in the projective plane, the first three of which do not lie on a line. Therefore we can pick a suitable coordinate system such that the points with homogeneous coordinates form a matrix M as below.

$$
\begin{array}{c}
1 \\ 2 \\ 3 \\ 4 \\ 5 \\ 6 \\ 7 \\ 8 \\ 9
\end{array}
\left(
\begin{array}{ccc}
1 & 0 & 0 \\
0 & 1 & 0 \\
0 & 0 & 1 \\
a & b & c \\
d & e & f \\
g & h & i \\
j & k & l \\
m & n & o \\
p & q & r
\end{array}
\right)
\qquad
\begin{array}{lll}
i\ m\ k\ p\ (bf - ce) & (1,4,5) \\
+\ i\ b\ k\ p\ (do - fm) & (2,5,8) \\
+\ i\ b\ d\ p\ (ln - ko) & (1,7,8) \\
+\ i\ b\ d\ l\ (mq - np) & (3,8,9) \\
+\ m\ b\ d\ l\ (hr - iq) & (1,6,9) \\
+\ m\ b\ r\ l\ (eg - dh) & (3,5,6) \\
+\ m\ b\ r\ e\ (ij - gl) & (2,6,7) \\
+\ m\ i\ r\ e\ (ak - bj) & (3,4,7) \\
+\ m\ i\ k\ e\ (cp - ar) = 0 & (2,4,9)
\end{array}
$$

The algebraic identity on the right of the matrix holds, since each left monomial cancels out with the right one in the line below (in a cyclic manner). Each factor in a line corresponds to a determinant of a 3×3-submatrix of M. We consider in particular in each summand the last factor to obtain the three-tuples of points (rows of the 3×3-submatrix) marked in the right column. Under a non-degeneracy condition which is given as a by-product of our polynomial (corresponding 4 determinants of 3×3-submatrices have to be non-zero), we have found the assertion of Pappus.

Theorem 10.3 *[Pappus] Eight collinear triples of points (from the above right column) imply the ninth triples of points to be collinear as well.*

10.5.2 The underlying oriented matroid result

Hyperline sequences of Pappus's configuration

```
([1], [[+2],[+4],[+5],[+9,+6],[+8,+7],[+3]])
([2], [[+1],[-3],[-7,-6],[-8,-5],[-9,-4]   ])
([3], [[+1],[+8,+9],[+7,+4],[+6,+5],[+2]    ])
([4], [[+1,-5],[-6], [-3,-7],[-8],[-9, +2]])
([5], [[+1,+4],[-3,-6],[-7],[-8, +2], [-9]])
([6], [[+1,+9],[+4],[-3,+5],[-7,+2], [-8] ])
([7], [[+1,+8],[+9],[-3,+4],[+5],[+6, +2] ])
([8], [[-7,+1],[-3,+9], +4, [+5, +2], [+6]])
([9], [[-6,+1],[-7],[-3,-8],[+4, +2], [+5]])
```

We consider now the oriented matroid aspect of the foregoing theorem which has led to this compact proof. Let us start with Pappus's configuration depicted in Figure 10.8. We have the corresponding hyperline sequences as a list of pairs: a rank 1 oriented matroid together with a rank 2 oriented matroid. We have written a corresponding uniform version as well. It corresponds to Figure 10.9.

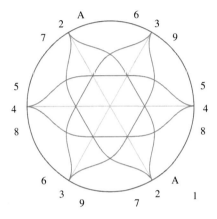

Figure 10.8   Pseudoline arrangement of Pappus's configuration

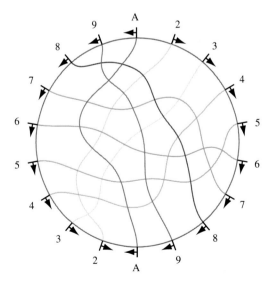

Figure 10.9   Pseudoline arrangement of uniform Pappus's configuration

Hyperline sequences of a uniform Pappus's configuration

```
1    +2, +4, +5, +6, +9, +8, +7, +3
2    +1, -3, -6, -7, -8, -5, -4, -9
3    +1, +8, +9, +4, +7, +6, +5, +2
4    +1, -5, -6, -7, -3, -8, +2, -9
5    +1, +4, -3, -6, -7, -8, +2, -9
6    +1, +4, -3, +5, +2, -7, -8, -9
7    +1, +8, +9, +4, -3, +5, +2, +6
8    -7, +1, -3, +9, +4, +5, +2, +6
9    +1, -7, -3, -8, +2, +4, +5, +6
```

The non-uniform version implies the following signed bases.

```
  sign [1,2,6] = sign [2,3,8] = sign [1,7,3]
= sign [2,3,9] = sign [3,1,4] = sign [2,3,5]
= sign [1,2,7] = sign [1,2,9] = sign [1,5,3] = +1
  sign [1,4,5] = sign [2,8,5] = sign [1,8,7]
= sign [3,8,9] = sign [1,6,9] = sign [3,6,5]
= sign [2,6,7] = sign [3,4,7] = sign [2,4,9] =  0
```

We can change the nine non-bases into positive bases without losing the oriented matroid property. We obtain the uniform oriented matroid. The corresponding matroid is trivial, all 3-tuples with three different elements form a basis.

We look again at the polynomial that we have encountered when we proved Pappus's theorem. We now replace the variables with determinants in the form of brackets.

$$
\begin{array}{c}
\begin{array}{l}
1 \\ 2 \\ 3 \\ 4 \\ 5 \\ 6 \\ 7 \\ 8 \\ 9
\end{array}
\left(
\begin{array}{ccc}
1 & 0 & 0 \\
0 & 1 & 0 \\
0 & 0 & 1 \\
a & b & c \\
d & e & f \\
g & h & i \\
j & k & l \\
m & n & o \\
p & q & r
\end{array}
\right)
\end{array}
\qquad
\begin{array}{l}
[1,2,6]\,[2,3,8]\,[1,7,3]\,[2,3,9]\,[1,4,5] \\
+\,[1,2,6]\,[3,1,4]\,[1,7,3]\,[2,3,9]\,[2,8,5] \\
+\,[1,2,6]\,[3,1,4]\,[2,3,5]\,[2,3,9]\,[1,8,7] \\
+\,[1,2,6]\,[3,1,4]\,[2,3,5]\,[1,2,7]\,[3,8,9] \\
+\,[2,3,8]\,[3,1,4]\,[2,3,5]\,[1,2,7]\,[1,6,9] \\
+\,[2,3,8]\,[3,1,4]\,[1,2,9]\,[1,2,7]\,[3,6,5] \\
+\,[2,3,8]\,[3,1,4]\,[1,2,9]\,[1,5,3]\,[2,6,7] \\
+\,[2,3,8]\,[1,2,6]\,[1,2,9]\,[1,5,3]\,[3,4,7] \\
+\,[2,3,8]\,[1,2,6]\,[1,7,3]\,[1,5,3]\,[2,4,9] = 0
\end{array}
$$

It turns out that the signs of all brackets lead to nine monomials that are all positive, a contradiction to the fact that in the case of a matrix, the polynomial has to be equal to zero. This implies that the uniform oriented matroid is not realizable.

The oriented matroid in terms of oriented bases is the following.

Oriented bases of nine-element oriented matroid

```
123+ 124+ 125+ 126+ 127+ 128+ 129+ 134- 135- 136- 137- 138-
139- 145+ 146+ 147+ 148+ 149+ 156+ 157+ 158+ 159+ 167+ 168+
169+ 178- 179- 189- 234+ 235+ 236+ 237+ 238+ 239+ 245- 246-
247- 248- 249+ 256- 257- 258- 259+ 267+ 268+ 269+ 278+ 279+
289+ 345+ 346+ 347+ 348- 349- 356- 357- 358- 359- 367- 368-
369- 378- 379- 389+ 456+ 457+ 458+ 459+ 467+ 468+ 469+ 478+
479+ 489+ 567+ 568+ 569+ 578+ 579+ 589+ 678+ 679+ 689+ 789+
```

The Pappus's configuration (not uniform) leads to the following oriented matroid, if we add an additional element. The chirotope is the following:

Oriented matroid with tenth element at infinity

```
10
3
+-++0--0+-0++0-0--0++++0--++0-++0+++0++++++------0++++++++++
00--++++++----0--0+0+++++++0--0--------0++++++++++++++++++++
```

By using the omawin program, we obtain the pseudoline arrangement of Figure 10.8. We did not mark the orientations, we have shown the reorientation class only.

Similar arguments lead to other proofs of projective incidence theorems like Desargue's theorem. The connection of a projective incidence theorem and a final polynomial argument for a non-realizable oriented matroid is a fruitful concept. The unpublished investigation of all non-realizable uniform rank 3 oriented matroids with ten elements, Bokowski, Laffaille, and Richter-Gebert, 1989, leads to a certain overview of projective incidence theorems that were not seen in this

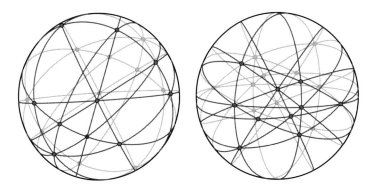

Figure 10.10 Configuration of two additional incidence theorems with ten elements

way before. Final polynomials for non-realizability proofs of oriented matroids have led to short proofs of projective incidence theorems.

In Figure 10.10 we show two additional examples of projective incidence theorems that occurred when non-realizable oriented matroids with ten elements were found. We just show their corresponding great circle arrangements.

10.6 Concept analysis

In Wille's school of Concept Analysis, see Ganter and Wille, 1999, mathematical aspects are used to support analysis and understanding of concept structures in many, especially non-mathematical, disciplines. In this section we show how oriented matroid techniques can be applied in this area. We pick an example from Bokowski and Kollewe, 1992.

We consider computer figures $1, 2, \ldots, 9$ as our objects. Each figure consists of a particular subset of segments $A, B, \ldots, G$ as shown below.

These segments are our items and the following incidence matrix tells us how these figures (objects) are composed of our items.

object	has			item			
	A	B	C	D	E	F	G
1	no	no	yes	no	no	yes	no
2	yes	no	yes	yes	yes	no	yes
3	yes	no	yes	yes	no	yes	yes
4	no	yes	yes	yes	no	yes	no
5	yes	yes	no	yes	no	yes	yes
6	yes	yes	no	yes	yes	yes	yes
7	yes	no	yes	no	no	yes	no
8	yes	yes	yes	yes	yes	yes	yes
9	yes	yes	yes	yes	no	yes	no

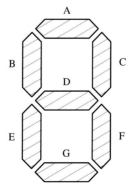

Figure 10.11 Segments for a computer figure eight

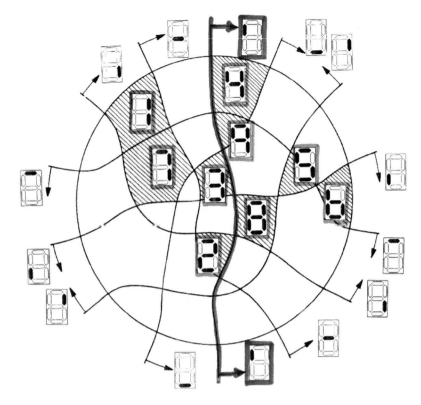

Figure 10.12   Pseudoline arrangement, a tool for concept analysis

Provided there exists an oriented pseudoline arrangement such that each item corresponds to a particular pseudoline and such that our objects can be marked as a 2-cell within this arrangement, we would have the incidence matrix in a nice geometrical picture that allows us to see the above incidences better. We compare two figures and we immediately see by how many items they differ.

For the example we can indeed find such a graphical representation. The resulting picture of Figure 10.12 was obtained by extending oriented matroids of rank 3 and discarding those that are not compatible with the incidence structure of the problem.

10.7   Triangular pseudoline arrangements

Let $\mathcal{A}$ be an arrangement of n pseudolines in the real projective plane and let $p_3(\mathcal{A})$ be the number of triangles of $\mathcal{A}$. We call a pseudoline arrangement triangular if $p_3(\mathcal{A}) = \frac{1}{3}n(n-1)$.

Grünbaum already studied the class of triangular pseudoline arrangements in Grünbaum, 1972, without using a name for it. Roudneff used the notion *tight* in Roudneff, 1986. In Bokowski, Roudneff, and Strempel, 1997, the term *p_3-maximal* *pseudoline arrangement* was used for that concept since in the case of a triangular pseudoline arrangement the number of triangles attain their general upper bound $n(n-1)/3$.

The reason for that is the following. If a pseudoline segment borders two triangles, the vertices of these triangles not lying on that segment have to coincide because of the intersection property of pseudolines. However, this in turn implies that two pseudolines have only two intersection points with other pseudolines leading to a contradiction for $n \geq 4$. In other words: two triangles cannot be adjacent to a pseudoline segment for $n \geq 4$. On the other hand, we have one triangle on each pseudoline segment of a triangular pseudoline arrangement, because of its definition, which leads to the maximal number $n(n-1)/3$ by counting incidences. Note that the number of pseudolines n must be even for $n \geq 4$.

Grünbaum has posed the following question. Are there infinitely many simple arrangements of straight lines that are triangular? Forge and Ramirez Alfonsin have proven this in the affirmative in Forge and Ramirez Alfonsin, 1998. From Figure 10.13 we can understand a decisive construction that they have used in their proof. Red curves have been added in the neighborhood of the curve, forming the line at infinity in the yellow and green triangular pseudoline arrangement with ten elements. As a result, we have obtained a new triangular pseudoline arrangement with 18 elements. A corresponding construction of adding additional pseudolines can be applied repeatedly. Thus, we obtain an infinite sequence of triangular pseudoline arrangements. There are other known constructions, starting with a triangular pseudoline arrangement, that obtain other triangular pseudoline arrangements with more elements. Such constructions are due to Harborth, 1985 and Roudneff, 1986. The article Bokowski, Roudneff, and Strempel, 1997 provides an algorithm for finding all

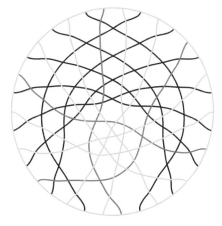

Figure 10.13 Method of Forge and Ramirez Alfonsin to obtain new triangular pseudoline arrangements from a given one

Figure 10.14   Pseudoline arrangement on a truncated cube, pottery model

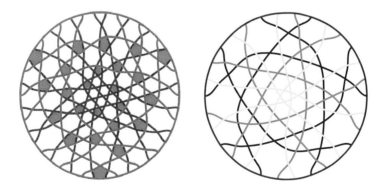

Figure 10.15   Triangular pseudoline arrangements

triangular pseudoline arrangements for a given number of elements. A corresponding functional program (chosen language: gofer, an antecedent of Haskell) was written by Bokowski and Jost. Results of this program lead to a complete list of existence of triangular pseudoline arrangements with up to 40 elements.

The example from Figure 10.14 was not known before that investigation. The symmetry of the pottery model indicates the symmetry of this triangular pseudoline arrangement.

The examples in Figure 10.15 are not only interesting as triangular pseudo-line arrangements. They can be viewed as special block designs, see Bokowski, Roudneff, and Strempel, 1997. Moreover, this article was the origin of a discovery that connects complete graph embeddings on 2-manifolds with a generalized concept of triangular pseudoline arrangements.

10.8 Oriented matroids are generalized Platonic solids

Properties of the Platonics, considered in a more general framework, lead to many mathematical concepts. Coxeter groups, regular maps, and graphs are such examples. The symmetry aspect plays a central role in the research exposition on abstract regular polytopes by McMullen and Schulte, 2002. It is a surprise that oriented matroids can also be viewed in this way. Drop the symmetry aspect and switch from the sphere to the projective plane. The combinatorial cell decomposition properties on the 2-sphere, obtained from the Platonics and carried over to the projective plane, provide us with a new definition for oriented matroids in rank 3. The concept of Petrie polygons, the key concept for regular maps, plays the central role.

For a given face-to-face cell-decomposition on a 2-manifold, we define a Petrie polygon to be a closed edge polygon with edges from the cell-decomposition such that

- for any pair of adjacent edges, we have an incident cell with both edges,
- a cell is never incident with three consecutive edges.

A Petrie polygon of a cell-decomposition has a Petrie polygon counterpart in its dual cell-decomposition. Thus a cell-decomposition such that all Petrie polygons have equal length k, implies the same property for the dual cell-decomposition. The dual pairs of Platonics have Petrie polygons of equal length. Moreover, any pair of these Petrie polygons meet in precisely one pair of antipodal edges.

If we consider a cell-decomposition of the projective plane, we can extend this via antipodal reflection to a covering of the 2-sphere. If we have in this case Petrie polygons of equal length and any pair of these Petrie polygons meet in precisely one pair of antipodal edges, we arrive at an object that corresponds to a rank 3 oriented matroid. For details, see Bokowski, Roudneff, and Strempel, 1997. As an example, study the following Figure 10.17. Pick a point in each spherical triangle of the left circle arrangement. Connect these points via arcs on the sphere if the triangles have a vertex in common. The resulting cell decomposition has Petrie polygons that correspond to the original great circles.

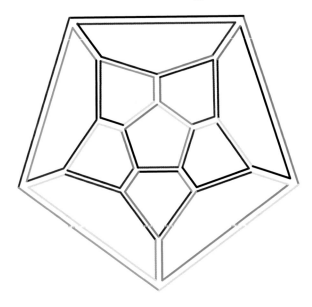

Figure 10.16 Petrie polygons of the dodecahedron

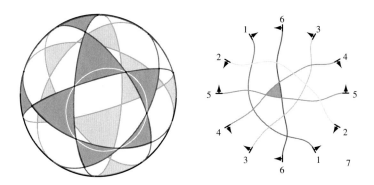

Figure 10.17 Six pseudolines represented on a sphere and in the projective plane

10.9 Curves on surfaces intersecting pairwise once

We exemplify a generalization of a pseudoline arrangement. We consider the example of a pseudoline configuration in the projective plane of Figure 10.18.

We are going to modify the fundamental polygon that defines the projective plane. Another way of gluing opposite edges of the fundamental polygon, the regular $2n$-gon, defines in general another surface. We keep the pseudoline intersection property of the projective plane. Thus we obtain n closed curves on a surface such that any two curves meet precisely once where they intersect.

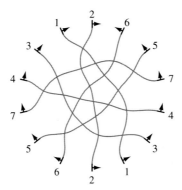

Figure 10.18   Closed curves on a surface intersecting pairwise once

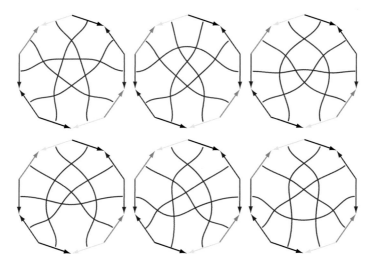

Figure 10.19   All six different equivalence classes of curve arrangements with five elements on an orientable surface of genus 2

Note that any orientable and non-orientable surface can be represented as a regular $2n$-gon (fundamental polygon) with a description of how the oriented edges of the fundamental polygon have to be glued.

As an example we consider an orientable surface of genus 2. The closed curves within the fundamental regions in Figure 10.19 show all combinatorial different curve arrangements with five elements on a surface of genus 2. A Haskell program – we do not include the program here – had been written to find all of them. The main idea is that we can use the data structure similar to that of hyperline sequences. We go along the oriented curves and we mark the way (by a sign) and the sequence of how the other curves cross. Looking at the same crossing point from the other curve requires a consistency for the data structure.

We first observe that there is essentially just one curve arrangement with four elements. We can write this in the following form.

```
e_4::[(Int,[Int])]
e_4=[(1,[ 2, 3, 4]),(2,[-1, 3, 4]),(3,[-1,-2, 4]),(4,[-1,-2,-3])]
```

Now we can check all inserting positions for the fifth element; we use all possible reorientations; we use the consistency condition and it turns out that the only examples are those from Figure 10.19. We have to check the surface as well. It is uniquely determined by the discs that are adjacent to the curve segments.

Let us assume that we have such a curve arrangement on an orientable surface. In this case we have the following theorem.

Theorem 10.4 *[Bokowski and Pisanski] Let g be the genus of an orientable surface S and let k be a natural number. There exists a curve arrangement of k closed simple curves on S that intersect pairwise precisely once if and only if $0 \leq k \leq 2g+1$.*

For the proof, which uses in one part co-homology theory, we refer the reader to the paper Bokowski and Pisanski, 2005,

The orientable case differs completely from the non-orientable one. Since each projective pseudoline arrangement can be generalized to a curve arrangement by attaching any number of handles to the projective plane, we conclude that for positive integers k and g' there exists a curve arrangement with k curves on a non-orientable surface of genus g'. Hence there are no upper bounds in the non-orientable case. We have curve arrangements in the non-orientable case of great profusion.

Theorem 10.5 *[Bokowski and Pisanski] For any non-orientable surface the number n of curves of a curve arrangement of simple closed curves that intersect pairwise precisely once is not bounded from above.*

Triangular curve arrangements

We generalize the concept of triangular pseudolines to that of our curve arrangements on surfaces that is, simple closed curves; each pair of curves has precisely one point of intersection.

Definition 10.6 *We call a curve arrangement on a surface triangular, if the cells in the cell decomposition of the surface are 2-colorable and if one component consists of triangles only.*

Complete graph embeddings on surfaces form a classical topic in topological graph theory with its Heawood map-coloring problem from 1890 and the Ringel–

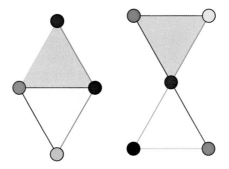

Figure 10.20 How we obtain locally the Petrie dual representation (left) from a triangulated map (right)

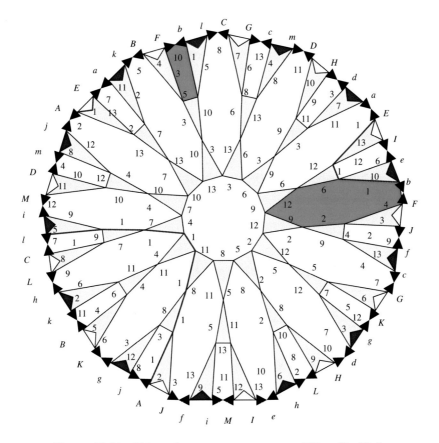

Figure 10.21 Triangular curve arrangement of Ringel's (S_{13})

Youngs solution from 1968, see Ringel and Youngs, 1968. For any integer $n \geq 7$ such that $g = (n-3)(n-4)/12 \in \mathbb{Z}$, there exists a triangulation of an orientable 2-manifold of genus g whose 1-skeleton is a complete graph with n vertices. All triangular complete graph embeddings on surfaces from Ringel, 1974, about the

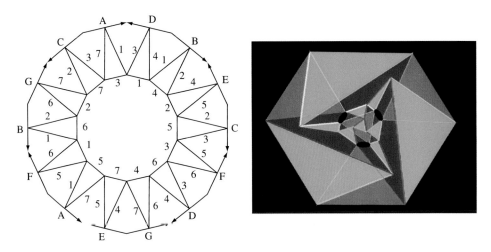

Figure 10.22   Seven simple closed curves on a surface of genus 3

map color theorem serve as examples of triangular curve arrangements due to the following transition.

We start with the triangular complete graph embedding and we form locally on a surface from a triangle its polar dual. We glue former adjacent triangles at their new vertices after we have changed their orientation. Figure 10.20 shows what we obtain locally. Adjacent cells change their orientation and vertex labels become edge labels and vice versa. The colors support our argument. Additional cells occur. They form together with all triangles a new surface, the cell decomposition of which is 2-colorable. We call it the *Petrie Dual representation*. The reason for this notation is that we can do the inverse operation for both color classes. The resulting surfaces are Petrie duals of each other.

We provide one example that corresponds to Ringel's (S_{13}) in Ringel, 1959, p. 71. It leads to a triangular curve arrangement with 13 curves. These curves intersect pairwise precisely once on an orientable surface of genus $g = 6$. We have depicted the triangular curve arrangement in Figure 10.21. As non-triangular cells we have three 13-gons and thirteen 9-gons.

We obtain another smaller example in Figure 10.22 where we have started with Möbius's torus.

10.10   Curves on surfaces intersecting at most once

We carry over the concepts of the previous section to any surface with a face-to-face cell decomposition derived from a simple closed curve arrangement. We look at a curve arrangement that we obtain from Klein's regular map, see for example

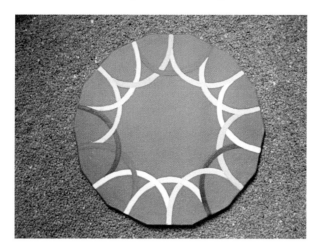

Figure 10.23 Pottery model of Fig. 10.2

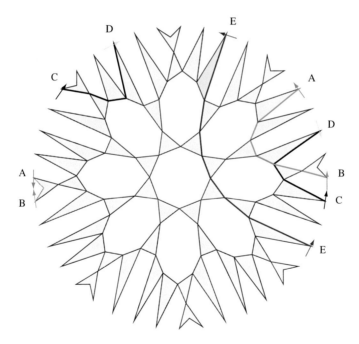

Figure 10.24 Petrie dual representation of Felix Klein's regular map $\{3,7\}_8$

Scholl, Schuermann, Wills, 2002, and a curve arrangement that we obtain from Dyck's regular map. In Figure 10.24 we have depicted the Petrie dual representation of the famous regular map $\{3,7\}_8$ of Felix Klein. Figures 10.25 and 10.26 show the Petrie dual representation of Dyck's regular map $\{3,8\}_6$.

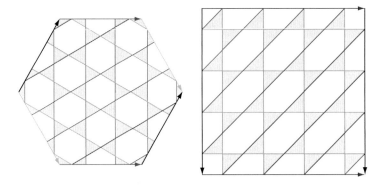

Figure 10.25 Petrie dual representation of Dyck's regular map $\{3, 8\}_6$

Figure 10.26 Petrie dual representation of Dyck's regular map $\{3, 8\}_6$ on the torus. Computer graphic of C. Séquin, Berkeley

The 2-colorable cell decomposition of the surface has only triangles in one component and only octagons in the other. Each of the 24 curves has seven intersection points. For properties of regular maps in a general framework see McMullen and Schulte, 2002, and the papers cited there.

10.11   On the Erdős–Szegeres conjecture

Goodman and Pollack have pointed out that a problem of Erdős–Szegeres has close links to the oriented matroid setting and their allowable sequence concept, see Goodman and Pollack, 1981. Let $g(m)$ be the smallest integer such that any planar configurations of $g(m)$ points, no three collinear, contains the vertices of a convex m-gon. From Erdős and Szegeres, 1935 and 1960, we know that

$$2^{m-2} + 1 \leq g(m) \leq \binom{2n-4}{n-2},$$

and we have the conjecture

$$g(m) = 2^{m-2} + 1.$$

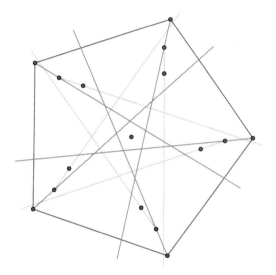

Figure 10.27   Example showing $17 \leq g(6)$ in the Erdős–Szegeres problem

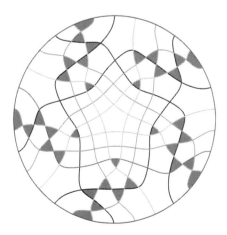

Figure 10.28   Pseudoline arrangement of the former example showing $17 \leq g(6)$ in the Erdős–Szegeres problem

Erdős and Szegeres also mention it is not feasible to test even the first case not yet resolved, namely that of $m = 6$. Let us look at an unpublished example of Bokowski that shows 17 to be a lower bound for $g(6)$, see Figure 10.27.

The proof uses a case analysis that we expect the reader to find on his/her own. The above picture is decisive for that. The symmetry helps to cut down the number of cases. For $m = 5$ we find a direct proof in the literature, an oriented matroid version approach to it was established in Engel, 1995. In Engel's Diplom thesis we find a detailed exposition of all cases for the above example. The author expects that a new computational approach with computers of our time can solve the Erdős–Szegeres conjecture for $m = 6$ on the oriented matroid level. The oriented matroid version of the Erdős–Szegeres problem is of course more general. However, if true, it does prove the original one.

Finally, we provide the pseudoline arrangement for the above example.

We observe that there are only triangular, quadrangular, and pentagonal cells in the pseudoline arrangement. This is of course no surprise. Moreover, a deletion of 11 elements cannot lead to a hexagonal cell.

11

Some intrinsic oriented matroid problems

We began to study oriented matroids by looking at equivalence classes of matrices. The theory of oriented matroids has now been seen as a mathematical field with many links to other areas. From a mathematical point of view, it is of interest to understand the oriented matroid concept without asking for its applications. This final chapter is devoted to some mathematical problems of the theory of oriented matroids.

11.1   Enumeration of small example classes

Non-realizability is a property of the reorientation class of an oriented matroid. The fact that we have non-realizable examples already with nine elements in rank 3 was already known to Levi in 1926. All uniform examples in rank 3 with less than nine vertices are realizable according to Goodman and Pollack, 1980. There is precisely one uniform example with nine elements in rank 3. This result was obtained first by Laffaille, 1989, and independently by Richter-Gebert, 1989. See also Gonzalez-Spinberg and Laffaille, 1989. The smallest non-realizable uniform examples in rank 4 have eight elements according to an unpublished result of Bokowski and Richter-Gebert. Here we provide a list of 24 chirotopes that form a complete list of representatives of all non-realizable reorientation classes of rank 4 with eight elements.

Representatives of all 24 non-realizable reorientation classes of oriented matroids with 8 elements in rank 4

```
+++++----- -----+++++ +-+----+-+ +++++----- -+-++++-+- ------++++ -+++++----+
+++++----- ----++++++ +++++----- -++------- -----++++++ +--++-++++ +++----++-
+++++----- ---+++++++ +++----+-+ +++++---+- -+-++++-+- ------++++ +++++-+---
++-+++---- --+++---+- +++++--+-- -----++-++ --+-++++++ -+++++++-+- -+++++-----
```

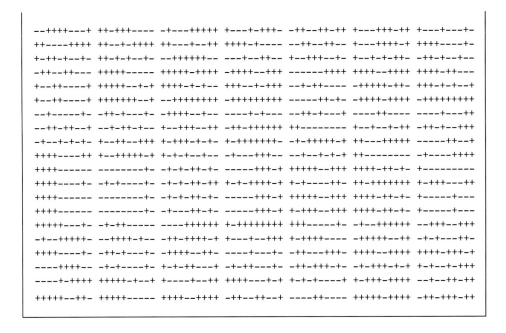

At present, we have oriented matroid databases by Aurenhammer and Finschi on the Internet. An old catalogue with examples of matroids can be found in Blackburn, Crapo, and Higgs, 1969.

11.2   The mutation problem

Some problems in convexity can be seen in a very large framework. Let us compare two major problem classes in convexity. Given the map that assigns to each non-empty compact convex set in Euclidean d-space all $d + 1$ Minkowski's quermassintegrals. Can we characterize, at least partially, the images in $\mathbb{R}^{d+1}$ that can occur? The classical isoperimetric inequality and many other theorems of Brunn–Minkowski theory are of this type.

Now we look at a combinatorial analog. Given the map that assigns to each convex polytope its f-vector, $f = (f_0, f_1, \ldots, f_{d-1})$, of the face lattice of the polytope. Can we characterize, at least partially, the images in $\mathbb{R}^{d+1}$ that can occur? Euler's classical formula – the alternating sum of all its components is constant – and many theorems in discrete convexity, like the upper bound theorem of McMullen, can be subsumed under this general heading.

So, it is natural to formulate a corresponding question for oriented matroids. Given the map that assigns to each rank r oriented matroid with n elements the list of combinatorial boundary types of the topes (cells of maximal dimension) of the big face lattice. Can we characterize, at least partially, the images that can occur?

Figure 11.1 A lamp inspired by the mutation problem, pottery model

Many results for the theory of pseudoline arrangements are partial answers to this very general question. A particular problem for uniform oriented matroids, due to Michel Las Vergnas, fits into this framework.

Problem 11.1 *[Las Vergnas] Does every uniform pseudo-hyperplane arrangement in rank r, r > 3, have at least one simplicial cell?*

Such a simplicial cell is referred to as a *mutation*. Changing the orientation of such a cell would lead to another oriented matroid. We define a graph for which the vertices are uniform oriented matroids of given rank r, $r > 3$, and given number n of elements. An edge occurs if and only if two oriented matroids differ by just one mutation. It is unknown whether this graph is connected.

Problem 11.1 sounds simple. However, we have no answer for it. Very often, an unsolved problem in mathematics leads to new methods in the corresponding mathematical field. So, from this point of view, it is useful to tackle this problem.

For an important example in this context (seven mutations and eight elements) it is useful to have a model to comprehend the full structure of the cell decomposition of the projective 3-space in this case. We have shown a topological model in Figure 11.2.

This oriented matroid arose as a contraction of the matroid polytope corresponding to Altshuler's sphere M_{963}^9. Another way to understand this cell decomposition shows the pottery model in Figure 6.7.

We describe this model according to the computer graphics in Figure 11.3 that was taken from Bokowski and Rohlfs, 2001. Figure 11.3 shows eight rank 3 contractions of it.

Figure 11.2   A first model in the city hall of Bremen on the occasion of a meeting of the German Mathematical Society 1990, photo B. Artmann

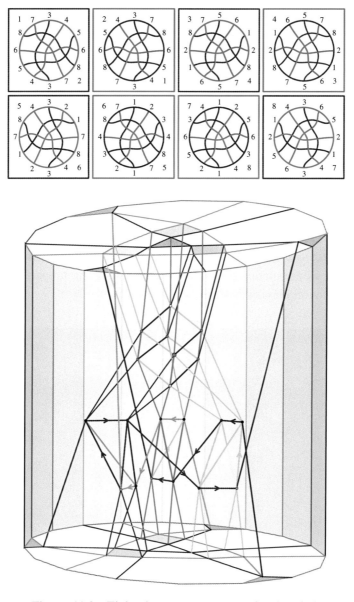

Figure 11.3 Eight elements, seven mutations, rank 4

The boundaries of the base element and the lid in Figure 11.3 are parallel elements. An additional lid parallel to the existing one on top has to be identified with the base element. The outer cylindrical part plays no mathematical role. The other six pseudoplanes intersect base and lid in pseudolines which indicate these pseudoplanes together with those pseudolines that are incident with pairs of those additional pseudoplanes.

Starting with this model, we can find an infinite sequence of extensions of this oriented matroid in which the number of mutations is smaller than the number of elements. The following model in Figure 11.4 shows a Folkman–Lawrence representation of it.

However, even the spatial impression of the model does not reveal a clear understanding of why the number of mutations is smaller than the number of elements.

In Richter–Gebert, 1993, we find a series of examples with a ratio of $(3n + 1)$ mutations and $4n$ elements, $n \geq 2$. The best result at present appeared in Bokowski and Rohlfs, 2001. Here we have oriented matroids with a ratio of $(5n + 4)$ mutations and $7n + 4$ elements. This article contains the smallest known rank 4 example in which one contraction has no adjacent mutation, see Figure 11.5.

All examples known to the author, in which the number of mutations is smaller than the number of elements, are extensions of the above example with eight elements and seven mutations. It is the only one with that property within all 24 rank 4 reorientation classes that are not realizable.

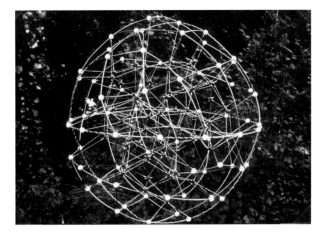

Figure 11.4　A model of P. Schuchert, nine elements, eight mutations. Complete model of projective three space (above) and detail (below)

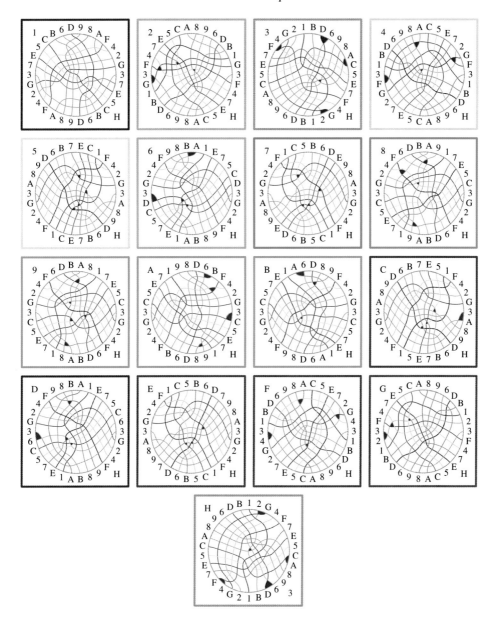

Figure 11.5   One contraction has no adjacent mutation, smallest known example

We mention in this context an infinite series of rank 3 examples of Bokowski (unpublished) with the property that all mutations are pairwise disjoint, see Figure 11.6. In Ljubic, Roudneff, and Sturmfels, 1989, another example with that property played a decisive role.

The following examples in Figure 11.7 were found by Roudneff (unpublished). There are constructions of his that lead to infinite classes of them. Perhaps by

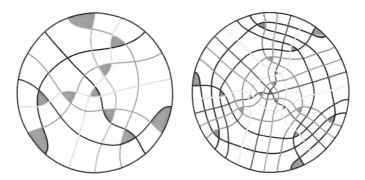

Figure 11.6   Beginning of an infinite series of examples with no adjacent mutations

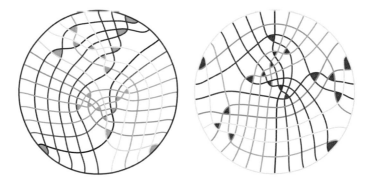

Figure 11.7   Additional examples of Roudneff, no adjacent mutations

starting with these examples as contractions, one can construct interesting rank 4 examples or even higher rank examples?

11.3   On the cube problem of Las Vergnas

Throughout this book ordinary matroids played no essential role. Many books on matroid theory do not cover oriented matroid aspects. One reason being that they were written before oriented matroid techniques had been developed. However, a profound understanding of ordinary matroid theory is useful for oriented matroid theory in many respects. As a consequence, we can recommend that the reader also study (ordinary) matroid theory via its many books. The theory of matroids was well established when the theory of oriented matroids started to become a discipline on its own. An essential problem is that of effectively characterizing matroids that are not orientable. Of course, we have many systems of axioms that would answer that question. However, from an algorithmic point of view, many systems of axioms are not applicable at all. This section deals with a particular problem that concerns matroids and their reorientation classes of oriented matroids.

The following problem of Las Vergnas concerns the n-dimensional cube. The problem has been solved up to dimension seven in Bokowski, Guedes de Oliveira, Thiemann, and Veloso, 1996.

When we start with the n-cube, we can determine its matroid and its oriented matroid. Now consider this matroid of the n-cube to be given. Las Vergnas has asked whether it determines uniquely an oriented matroid up to reorientation (and sign reversal). In other words:

Problem 11.2 *[Las Vergnas] When a rank $n + 1$ oriented matroid has as its underlying matroid that of the n-cube, can it always be transformed via reorientations of some of its elements into that of the n-cube?*

An interesting outcome of this investigation was the symmetry group of the matroid of the unit n-cube. Consider the Coxeter group of the unit $(n + 1)$-cube, that is the group of rigid motions that keeps the unit $(n + 1)$-cube fixed. When is taken the subgroup G in which opposite vertices with distance $\sqrt{n+1}$ keep their distance is taken, it also acts on the n-cube. This restriction of G on the unit n-cube is a proper super group of the Coxeter group of the n-cube and it is the automorphism group of the matroid of the n-cube according to Bokowski, Guedes de Oliveira, Thiemann, and Veloso, 1996.

Appendix A   A Haskell primer

We provide neccessary explanations of our Haskell program. For more aspects of this academic freeware, we refer the reader to the URL address `http://www.Haskell.org`. See also Hall and Donnell, 2000, where a short introduction to Haskell can be found.

A.1   About Haskell B. Curry

We find a short description of Haskell B. Curry within the *Mathematics Genealogy Project*, see `http://www.genealogy.ams.org`.

Haskell Bruce Curry was born in 1900 in Millis, Massachusetts, USA. He died in 1982 in State College, Pennsylvania, USA. His Ph.D. dissertation in Göttingen, Germany, was supervised by David Hilbert. It was entitled *Grundlagen der kombinatorischen Logik*.

Figure A.1   Haskell Brooks Curry

J. J. O'Connor and E. F. Robertson write about Curry

He taught at Harvard, Princeton, then for 35 years at Pennsylvania State University. During World War II Curry researched in applied physics at John Hopkins University. In 1966 he accepted the chair of mathematics at Amsterdam. Curry's main work was in mathematical logic with particular interest in the theory of formal systems and processes. He formulated a logical calculus using inferential rules. His works include Combinatory Logic *(1958) (with Robert Feys) and* Foundations Of Mathematical Logic *(1963).*

The work of Haskell B. Curry was honored by using his name for *Currying*, a special notation for a particular interpretation of mathematical functions with more than one variable. This can be explained by using a function f with two arguments a and b with image c, $f : (a, b) \mapsto c$. Currying stands for interpreting f as a function that assigns to the argument a the function $g : b \mapsto c$ with pre-image b and image c, $f : a \mapsto (b \mapsto c)$.

Moreover, Curry's first name Haskell was used for the de facto standard functional language with the lazy evaluation property, Haskell 98.

A.2 First Haskell functions and introduction

We provide a few Haskell functions that seem to be self-explanatory. Our intention is to increase the reader's motivation to study some syntax rules of the Haskell language later.

A first example concerns Pythagorean triples with components between 1 and n. The following function yields these triples for given n.

Pythagorean triples

```
pythtriple::Integer->[[Integer]]
pythtriple n = [[a,b,c]|a<-s,b<-s,c<-s, a*a+b*b==c*c]
               where s = [1..n]
```

Haskell syntax is close to mathematics. Square brackets are used for lists. The function is already the program. This gift of a development over decades of research within a branch of computer science is to our benefit.

A second example uses n factorial for natural numbers n. The following recursive function is at the same time the program.

n factorial

```
fak:: Integer->Integer
fak 0 = 1
fak n = n * (fak (n-1))
```

Haskell is very useful for recursive functions.

We use a third example. For a natural number r and a list of integer numbers, we determine all r-tuples that are sorted as in the list. For $r = 1$ the result is a list of one-element lists. For a given list with r elements, the result is just a list containing the given list. Otherwise, we first use the head element x of the given list and an $(r-1)$-tuple from the remaining list to form all r-tuples with the head element at the beginning. We finally append all r-tuples not containing the head element e, that is all r-tuples from the tail of the given list when the head element has been deleted. We can analyze the recursive structure, either the number of elements of the tuples has become smaller or the list has one element less.

All r-element tuples from a list

```
tup::Int->[Int]->[[Int]]
tup r l
 |r == 1      = [[el]|el<-l]
 |length l==r = [l]
 |True=(map([head l]++)(tup(r-1)(tail l)))++tup r(tail l)
```

You can see that ++ denotes the concatenation of two lists. The map function links together the one element list with the head element of l with all $(r-1)$-tuples that are the result of tup (r-1) (tail l).

Computer scientists and engineers from software engineering often first recommend that a problem is implemented in a functional programming language before the problem is to be implemented in another language like C or Java. The whole time up to the final implementation with possible changes of the problem and all searches for faults is in general much shorter. Programs are easier to comprehend after years, loop parameters do not appear explicitly in Haskell. Programmers that have worked both with imperative languages like Fortran, Pascal, C, C++, and Java and with a functional language like Haskell, report that only after their own experience with Haskell do they really appreciate all the advantages of the Haskell language, for example its compact code and its short time for fault detection.

We cite about the history of Haskell from the preface of Peyton and Jones, 2003.

In September of 1987 a meeting was held at the conference on Functional Programming Languages and Computer Architecture (FPCA '87) in Portland, Oregon, to discuss an unfortunate situation in the functional programming community: there had come into being more than a dozen non-strict, purely functional programming languages, all similar in expressive power and semantic underpinnings. There was a strong consensus at this meeting that more widespread use of this class of functional languages was being hampered by the lack of a common language. It was decided that a committee should be formed to design such a language, providing faster communication of new ideas, a stable foundation for real applications development, and a vehicle through which others would be encouraged to use functional languages. This book describes the result of that committee's efforts: a purely functional programming language called Haskell, named after the logician Haskell B. Curry whose work provides the logical basis for much of ours.

Goals

The committee's primary goal was to design a language that satisfied these constraints:

1. *It should be suitable for teaching, research and applications, including building large systems.*
2. *It should be completely described via the publication of a formal syntax and semantics.*
3. *It should be freely available. Anyone should be permitted to implement the language and distribute it to whomever they please.*
4. *It should be based on ideas that enjoy a wide consensus.*
5. *It should reduce unnecessary diversity in functional programming languages.*

...Haskell 98 provides a stable point of reference so that those who wish to write text books, or use Haskell for teaching, can do so in the knowledge that Haskell 98 will continue to exist.

The next example function defines prime numbers by using the sieve of Erastostenes. We take an ordered list of natural numbers. We take the first element and we delete all multiples of this first element in the list. We continue this process repeatedly. This can be described via some Haskell code.

Sieve of Erastostenes

```
ps=sieve[2..] where sieve(p:ns)=p:sieve[n|n<-ns,n'mod'p>0]
```

We obtain the first 15 prime numbers via `take 15 ps`.
The algorithm *quicksort* that sorts a list of integers is very short.

Quicksort algorithm

```
qsort::[Int]->[Int]
qsort [] = []
qsort (x:xs) = (qsort smaller)++[x]++(qsort greater)
          where
          smaller = [el|el<-xs, el<x]
          greater = [el|el<-xs, el>x]
```

Writing down precisely the mathematical idea and using the syntax rules of Haskell leads to the interpretable program. We close this section with Fibonacci numbers and a corresponding Haskell program for them.

Fibonacci numbers

```
fib::Int->Int
fib 0 = 0
fib 1 = 1
fib n = fib(n-1) + fib(n-2)
```

Every Haskell program can be compiled into a corresponding C-program, for example by using the ghc-compiler. This allows running time to be very much improved. For many problems, especially for those used for teaching, running time arguments are secondary. Having a first running program within reasonable time is decisive. Should running time and storage be improved later, we can use an appropriate other language.

Comments can be written after – – in a line or within brackets of the form {–...–}.

Remarks about module types can be found on page 303.

For the use of the sign "@" see the example on page 14.

Appendix B Software, Haskell functions, and examples

We have sorted the Haskell functions from all chapters and we provide references for those sections and pages where a more detailed description of corresponding functions have been given.

B.1 Software about oriented matroids

Remarks about software aspects have been made at several places within this book.

We have stressed functional programming and especially Haskell for working with oriented matroids. All used Haskell functions from this book are available via the URL of the author: `http://juergen.bokowski.de`.

The omawin program has evolved from a German Research Foundation project of the author. Many research students, especially J. Richter-Gebert, P. Schuchert, and others, have created this program in joint work with the author. It provides a tool to convert a rank 3 chirotope automatically into its Folkman–Lawrence representation and it converts a rank 4 chirotope into all its rank 3 contractions. It is available via the above URL address of the author.

A C++-program of the author for extending oriented matroids in rank 4 was decisive for an important application of oriented matroid theory in computational synthetic geometry. It can be used for many additional 3-dimensional problems. The program is available via the above URL address of the author.

For large problem classes, additional software has to be written. Many people have written software about oriented matroids. The author wishes to mention especially the programs of David Bremner, Lukas Finshi, and Ralf Gugisch concerning extensions of oriented matroids.

B.2   Used basic Haskell functions

We list important Haskell functions from the file `Prelude.hs` and from a standard library of list functions in the file `List.hs`. These functions should be clear from the function name with its type below, its verbal description and some examples on the right.

`(:)` `a->[a]->[a]`	Add a single element to the front of a list. `2:[4,5]` leads to `[2,4,5]`
`(++)` `[a]->[a]->[a]`	Join two lists together. `[2,4,5]++[3,5,2]` leads to `[2,4,5,3,5,2]`
`(!!)` `[a]->Int->a`	`xs!!n` returns n'th element of xs, starting at the beginning and counting from 0. `[2,4,5]!!2` leads to `5`
`concat` `[[a]]-> [a]`	Concatenate list of lists into single list. `concat[[3,4],[],[3,6]]` leads to `[3,4,3,6]`
`length` `[a]->Int`	The length of the list. `length [2,4,5]` leads to `3`
`head` `[a]->a`	The first element of the list. `head [2,4,5]` leads to `2`
`tail` `[a]->[a]`	All but the first element of the list. `tail [2,4,5]` leads to `[4,5]`
`take` `Int->[a]->[a]`	Take n elements from the front of the list. `take 2 [2,4,5,6]` leads to `[2,4]`
`drop` `Int->[a]->[a]`	Drop n elements from the front of the list. `drop 2 [2,4,5,6]` leads to `[5,6]`
`reverse` `[a]->[a]`	Reverse the order of the elements. `reverse [2,4,5]` leads to `[5,4,2]`
`zip` `[a]->[b]->[(a,b)]`	Take a pair of lists into a list of pairs. `zip [2,4] [3,5,6]` leads to `[(2,3),(4,5)]`
`and` `[Bool]->Bool`	The conjunction of a list of Boolean. `and [False,True,True]` leads to `False`

```
or                      The disjunction of a list of Boolean.
[Bool]->Bool            or [False,True,False]  leads to   True


sort                    Sorts the list.
[Int]->[Int]            sort [2,6,4,3,5]  leads to  [2,3,4,5,6]


union                   The union of two lists
[a]->[a]->[a]           union [1,3,5] [3,4,5]  leads to  [1,3,5,4]


fst                     Returns the first element of a pair
(a,b) -> a              fst (4,2) leads to 4


snd                     Returns the second element of a pair
(a,b) -> b              snd (4,2) leads to 2


(\\)                    Setminus, removes elements of second list
[a]->[a]->[a]           [2,4,6,8]\\[3,4,5] leads to [2,6,8]


map                     A function (a->b) applied to each element
(a->b)->[a]->[b]        a of a list [a].
                        map (2+) [1,3,5] leads to [3,5,7]
                        Another way to write this:
                        map (\i->2+i) [1,3,5]


odd                     Returns True if the integer is odd.
Integer -> Bool         odd 3 leads to True, odd 8 leads to False


even                    Returns True if the integer is even.
Integer -> Bool         even 3 leads to False, even 8 leads to True


elem                    Returns True if a belongs to list.
a -> [a] -> Bool        elem 3 [2,3] or 3 'elem' [2,3] returns True


notElem                 3 'notElem' [2,3] returns False
a -> [a] -> Bool


replicate               Returns a list which contains the element a
Integer->a->[a]         as often as the integer value tells us.
                        replicate 3 7 leads to [7,7,7]


elemIndices             For element a, we get the list of positions
a->[a]->[Int]           where a occurs within the list,
                        elemIndices 3 [2,3,4,1,2,3] leads to [1,5]
```

```
(==)                equal?          4 == 2*2   returns True
a -> a -> Bool


(/=)                not equal?    5 /= 4      returns True
a -> a -> Bool


filter              filter odd [1..8] returns [1,3,5,7]
(a->Bool)->[a]->[a]


div                 div 10 5    or    10 'div' 5   returns 2
Integer->Integer
        ->Integer
nub                 Removes dublicate elements from a list
[a]->[a]            nub [1,3,1,4,3,3] leads to [1,3,4]
sum                 the usual sum
[Integer] -> Integer
product             the usual product
[Integer] -> Integer
splitAt             Split a list at a given position.
Int->[a]->([a],[a])
                    splitAt 2 [2,4,5]  leads to  ([2,4],[5])
```

These functions indicate: we use just a fraction of Haskell's potentiality.

```
(:) (++) (!!) (\\) (==) (/=)
and concat fst div drop elem elemIndices even
filter head length map notElem nub odd or product
replicate reverse snd sort splitAt sum tail take union zip
```

We recommend for further reading Hudak, Peterson, and Fasel, 1997, Thompson, 1999.

The Haskell syntax allows us to define an expression within another part of the function in the form:

$$\texttt{let}\quad \texttt{exp} = f(x, y, \ldots)\quad \texttt{in}$$

In this case we define locally the variable `exp` in terms of other variables `x,y`.

We can define a function not only for a specific type but for a whole class of types, for instance for those that allow us to use the equality sign. In this case we can write the definition of the function, for instance, as `(Eq a)=>a->[a]->a`. We have sometimes used this construction.

Table B.1 *Types used in our Haskell functions*

```
type Es =[[Int]]                          -- edges
type GR =(Vs,Es)                          -- graph
type MA =[[Integer]]                      -- matrix
type OB =(Tu,Or)                          -- oriented base
type OE =(Vs,Int)                         -- oriented edge
type OF =[Int]->[Int]->Ordering           -- order function, rank~2
type OM1= [Int]                           -- rank1 oriented matroid
type OM2=[[Int]]                          -- rank2 oriented matroid
type OM3=[(OM1,OM2)]                       -- rank3 oriented matroid
type OM5=[([Int],OM2)]                     -- rank5 matroid polytope
type Or =  Int                            -- orientation
type Star=(Int,[Int])                     -- star  (vertex,link)
type Tu = [Int]                           -- tuple of elements
type Vs = [Int]                           -- vertices
type Ngon = [Int]                         -- hyperline, Pol., r = 5
```

Table B.2 *Haskell functions*

Function of Type	
Short explanation	Page
`adjEl::Star->Star->[Int]`	153
star->star->adjacents elements	
`after::(Eq a)=>a->[a]->a`	153
used in adjEl	
`all5::[[Or]]`	88
candidates of chirotopes, rank 3, 5 elements	
`allLinks::[[Int]]->[(Int,[[Int]])]`	234
`altOM::Int->Int->[OB]`	15
r->n->alternating oriented matroid, n elements, rank r	
`baseOfHp::[Int]->[Int]`	126
used in chi2Coc	
`before::(Eq a)=>a->[a]->a`	153
used in adjEl	
`bracket2Index::([Int],Int)->Int`	211
index of a bracket	
`cand::[[([Int],OM2)]]`	88
candidates of hyperline sequences, rank 3, 5 elements	
`cComp3::Int->[OB]->[Int]->Int`	57
used in chi2C for finding circuits of a chirotope	
`cComp4::Int->[OB]->[Int]->Int`	57
used in chi2C for finding circuits of a chirotope	

Table B.2   *(cont.)*

Function of Type Short explanation	Page
cEdge::[Int]->GR->[[OE]] [z,a]->graph->circuits with edge ([z,a],1)	60
change::[Int]->[Char] signs->characters for signs	133
changeSign::[Int]->OB->OB	117
checkOM3Pair::[(Tu,OM2)]->[(Tu,OM2)]->[([Int],[Int],Bool)] equivalence check	91
chi2C::[OB]->[[Int]]->[[Int]]->[[Int]] chi > [dop] > [] > circuits of chi	57
chi2Coc::[OB]->[[Int]] chi->cocircuits	126
chi2Facets::[OB]->[[Int]] chi->facets of chi	135
chi2HypR2::[OB]->OM2 chi->hyperlines, rank 2	79, 172
chi2HypR3::[OB]->[([Int],OM2)] chi->hyperlines, rank 3	82
chi2HypR5::[OB]->[OM5] chi->hyperlines, rank 5	172
chi2ordf::[OB]->([Int]->[Int]->Ordering) chi->ordering function in hyperline	78, 172
chirotope::[Int] an example 4-chirotope with 10 elements	208
cl::[Int]->[OB]->[Int] set->chi->closure of set (chi)	24, 172
clHP::[Int]->[OB]->[Int] hyperplane->chi->closure of hyperplane in chi	126
completeStars::Star->[[Int]]->[Star] complete possible stars	151
consR5::[Int]->[Int]->[OB]->[OB] sign consequences from GP relations	148
consistentStar::Star->[Star]->Bool star->partial 2-manifold->star consistent?	153
consistentStarPair::Star->[Star]->Bool used in consistentStar	153
contra::[Int]->[OB]->Bool 4 elements->chi->GP contradiction?	150
contraGP::Int->[OB]->Bool n->chi->any GP contradiction?	150
cs::[[Int]] cocircuit result, function for Petersen graph investigation	64

Table B.2 *(cont.)*

Function of Type Short explanation	Page
`ctrElChi::Int->[OB]->[OB]` `el->chi->contracted chi at el`	23, 170
`ctrSetChi::[Int]->[OB]->[OB]` `set->chi->contracted chi at set`	23, 170
`ctrST::Int->OB->[OB]` `used in ctrElChi`	170
`cube2Chi::Int->[OB]` `d->chirotope of d-cube`	119
`cube2MA::Int->MA` `d->matrix of d-cube`	119
`cube2MAHom::Int->MA` `d->matrix (homogeneous) of d-cube`	119
`cube3::[[Int]]` `homogeneous coordinates of 3-cube`	136
`dCr::[Int]` `result, a dual 6-chirotope with 10 elements`	209
`dC::Int->Int->[Int]->[Int]` `r->n->chi->dual chirotope of chi`	43
`delElChi::Int->[OB]->[OB]` `el->chi->deletes el in chi`	22, 169
`delSetChi::[Int]->[OB]->[OB]` `set->chi->deletes set in chi`	22, 169
`det::MA->Integer` `matrix->determinant of matrix`	73
`dets::[[Int]]->MA->[Integer]` `rsets->matrix->(r x r)-sub-determinants`	73
`e::[[(Int,[[Int]])]]->[[(Int,[[Int]])]]` `extending the star structure`	252
`e_4::[(Int,[Int])]` `curve arrangement with four elements`	278
`edges::[[Int]]->[[Int]]` `function for Szilassi's polyhedron`	225
`edges6gon::[Int]->[[Int]]`	225
`edgeTest::[[Int]]->Bool` `used within pseudoManifoldTest`	233
`eInClSet::Int->[Int]->[OB]->Bool` `e->set->chi->e in closure of set of chi?`	24
`eInHl::Int->Ngon->[Int]->[(Tu,Or)]->Bool` `used in cl`	172
`es2gr::[[Int]]->GR->GR` `edges->graph->subgraph without edges`	63

Table B.2 *(cont.)*

Function of Type Short explanation	Page

`ex::[(Int,[[Int]])]`
  list of (facet number, list of oriented triangles)
  of 3-sphere of Gévay
	194

`ext::Int->Int->Int->OM2->OM2` (two versions) used in inRow	166, 177
`extsmani::[Star]->[Int]->[Star]->[[Star]]` extensions of partial 2-manifolds	155
`extMP::Int->Int->[Or]->[([OM5],[Or])]` extensions of matroid polytopes, non-uniform	176
`extMPu::Int->Int->[Or]->[[Or]]` extensions of matroid polytopes, uniform case	167
`f::[Int]->[Int]->Ordering` used in chi2ordf	172
`facetsA::[[Int]]` facets of a 3-sphere of Altshuler	145
`facetsB::[[Int]]` facets of a 3-sphere of Brueckner	145
`facetsK::[[Int]]` facets of a 3-sphere of Kleinschmidt	143
`facetsP::[[Int]]` facets of a 3-sphere of pottery model of Figure 5.6	141
`facetsS::[[Int]]` facets of a 5-sphere of Shemer	207
`fak::Integer->Integer` Haskell primer function	296
`fib::Int->Int` Haskell primer function	298
`findNorm::[[Int]]->[[Int]]` for finding normalized hyperline in rank 3	90
`findRankInd::[Int]->[OB]->(Int,[Int])` set->chi->(k,ind), (rank of set, k-base in set)	23, 170
`fSort::Int->Int->Ordering`	174
`g::[(Int,Int)]->[((Int,Int),(Int,Int))]`	211
`genusManifold::[(Int,[Int])]->Int` stars->genus of triangulated 2-manifold	152
`genSets::[OB]->[[Int]]` chi->preparation 1 of hyperline data	77, 173
`glue::[OB]->[[Int]]->[[Int]]` chi->hyperline data->hyperline data	77, 173
`gpCheckR2::[Int]->[OB]->Bool` consistency check in rank 2	116

Table B.2 *(cont.)*

Function of Type Short explanation	Page
`gpRel2Products::([Int],[Int])->[([Int],Int)]`	211
`gpRelations::[([Int],[Int])]`	211
preparation of GP-relations, 10 elements	
`gr2C::GR->[[OE]]`	60
graph->circuits of graph	
`graphComplete::Int->([Int],[[Int]])`	58
n->complete graph with n vertices	
`graphP::GR`	60
Petersen graph (an oriented version)	
`hexagonsSzilassi::[[Int]]`	225
all hexagons of Szilassi's torus	
`hl2Pair::Ngon->[(Tu,Or)]->OM5`	174
ngon->chi->pair of hyperline with its sequence	
`hlBds::[[Int]]->[([Int],[Int])]`	147
simplicial facets->hyperline bounds	
`homCoord::MA`	81
homogeneous coordinates of example matrix	
`hyp2ChiR2::OM2->[OB]`	76
rank 2 hyperlines->chirotope	
`hyp2ChiR3::[([Int],OM2)]->[OB]`	82
rank 3 hyperlines->chirotope	
`hyp2ChiR5::[(Tu,OM2)]->[OB]`	171
rank 5 hyperlines->chirotope	
`hyp2Matroid::[([Int],OM2)]->[[Int]]`	24
rank 3 hyperlines->underlying matroid	
`hyperlines::[([Int],OM2)]`	24
hyperline data structure of example matrix M	
`hypex::[([Int],OM2)]`	81
hyperline sequences of an example	
`inAll::Int->Int->Int->([OM5],[Or])->[([OM5],[Or])]`	167, 176
all matroid polytopes extensions	
`inHl::Int->Int->([OM5],[Or])->[([OM5],[Or])]`	177
used in cl	
`inOM2::Int->Int->([OM5],[Or])->[([OM5],[Or])]`	178
inserting in rank 2 oriented matroid	
`inRow::Int->Int->([OM5],[Or])->([OM5],[Or])`	166, 177
inserting an element in a row, uniform case	
`ins::[Int]->(Int,[[Int]])->(Int,[[Int]])`	252
inserting a line segment in a star	
`isM::[[Int]]->Bool`	13
potential bases of a matroid->is it a matroid?	

Table B.2 *(cont.)*

Function of Type Short explanation	Page
isSphere::[(Int,[Int])]->Bool stars->do they form a 2-sphere?	152
isSubset::[Int]->[Int]->Bool set1->set2->set1 contained in set2?	55
isTorus::[(Int,[Int])]->Bool stars->do they form a torus?	152
itr::[Int]->(Int,[[Int]])->(Int,[[Int]]) insert triangle	251
kleinBottle::[[Int]] triangle list of a Klein bottle	233
known::Int->Bool used in consR5	148
kTermGPrels::Int->Int->Int->[[(([Int],[Int])]] k->r->n->k-term GP-relations	45
lFarkas::Int->Int->GR->(Es,Bool) vertex 1->vertex 2->graph->(path,True) or (cut,False)	66
linkGen::Int->[[Int]]->Int->[Int]->[[Int]]->(Int,[[Int]])	234
linkNumberTest::[[Int]]->Bool	235
listOfNewSigns::Int->[OB]->[OB]	149
lns::[[Int]] lines of 40_4 configuration	264
m::MA main sample matrix of Chapter 1 and Chapter 2	73
m1exp::Int->Int->[Int]->Int exponent function used in function dC	43
m2Chi::MA->[OB] matrix->chirotope of matrix	74
m416::MA coordinates of Altshuler's sphere 416	137

man::[[Int]]->[[Star]]
   triangle list->list of orientable triangulated
   2-manifolds
	156

manifolds::[Star]->[[Star]] star list->orientable 2-manifolds	154
manfoldTest::[[Int]]->Bool	234
mex::MA coordinates of example matrix	81
minDep2::MA->[[Int]] matrix->2-element minimal dependent sets	55
minDep3::MA->[[Int]] matrix->3-element minimal dependent sets	55

Table B.2 *(cont.)*

Function of Type Short explanation	Page
`minDep4::MA->[[Int]]`	55
matrix->4-element minimal dependent sets	
`modifiedSign::([Int],Int)->Int`	211
`mutOfChi::[Int]->[OB]->Bool`	117
for finding a mutation in a chirotope, rank 4	
`newOrEmpty::Int->[Or]->[(Tu,Or)]->[[Or]]`	166, 178
used in inRow	
`nGonSort::[Int]->OM2->[Int]`	174
`nOfComp::GR->Int`	62
graph->number of components of graph	
`norm::OB->OB`	21, 169
oriented base->normalized oriented base	
`normSeq::(Tu,OM2)->(Tu,OM2)`	90
for finding normalized hyperline sequence	
`normSeqs::[(Tu,OM2)]->[(Tu,OM2)]`	90
for finding normalized hyperline sequence	
`ns::[Int]`	64
function for Petersen graph investigation	
`nSO::[[Int]]->[([Int],[Int])]->[OB]->[OB]`	147
generates necessary simplex orientations for a 3-sphere	
`nStars::[[Int]]->Int->[Star]`	151
used in posStars	
`nstars::Int->[Star]->[Star]`	154
vertex->list of stars->stars with vertex	
`on_ln::[Int]->[Int]->Bool`	264
line condition for point pairs in a 40_4 configuration	
`pairs::OM2->OM2`	177
`pathCut::Int->Int->[Int]->GR->(Es,Bool)`	66
part of proof of Farkas's lemma for graphs	
`paths::Int->Int->[OE]->GR->[[OE]]`	60
i->z->path->graph->[path & path from i to z]	
`pb::[[Int]]`	13
potential bases of a matroid	
`perm::[Int]->(Int->Int)`	87
list->(i->list!!(i-1))	
`permSet::[Int]->[[Int]]`	87
list->list of all permutations of list	
`pinchedSphere::[[Int]]`	238
triangle list of pinched sphere	

Table B.2   *(cont.)*

Function of Type Short explanation	Page
plm::Int->Int	43
used in function dC for finding a dual chirotope	
posStar::Star->[Star]->[Star]	154
star->star list->add star to list when consistent	
posStars::[[Int]]->[[Star]]	151
list of triangles->lists of possible stars	
preHypR2::[(Tu,Or)]->[[Int]]	173
prepts::[[Int]]	263
preliminary points, input for a 40_4-configuration	
prod::[Char]->[Char]->Int	65, 264
circuit->cocircuit->orthogonal property	
prod::[Int]->[Int]->Int	264
an abstract product function, for point line configuration	
pseudoManifoldTest::[[Int]]->Bool	233
pythtriple::Integer->[[Integer]]	295
Haskell primer function	
qsort::[Int]->[Int]	297
Haskell primer function	
r::Int->[Int]	63
k->remaining number of components (Petersen graph)	
r2ex::[OB]	77
oriented bases in rank 2 example	
rankRm1::[OB]->[[Int]]	126
chi->all (rank-1)-flats of chi	
reduce::[[Int]]->[[Int]]	263
reduction of preliminary points of 40_4 configuration	
rel::(Int->Int)->[Int]->[Int]	89
used for relabeling in relab	
relab::(Int->Int)->(Tu,OM2)->(Tu,OM2)	89
used for relabeling in rank 3	
relabelChi::(Int->Int)->[OB]->[OB]	87
permutation->chi->relabeling of chi	
relabelHypR3::(Int->Int)->[(Tu,OM2)]->[(Tu,OM2)]	89
relabeling of hyperlines in rank 3	
relabelST::(Int->Int)->OB->OB	87
permutation->signed tupel->relabeling of signed tupel	
reorChiTuple::Int->OB->OB	86
k->base->reoriented base at k	
reorElChi::Int->[OB]->[OB]	86
el->chi->reoriented chi at el	

Table B.2   *(cont.)*

Function of Type Short explanation	Page
`reorElR3::Int->[(Tu,OM2)]->[(Tu,OM2)]` for reorSetR3	90
`reorLists::Int->[[Int]]` n->concat (map(i->tuples i n) [0..n])	86
`reorR1::Int->[Int]->[Int]` for reorRowR3	90
`reorRowR3::Int->(Tu,OM2)->(Tu,OM2)` for reorElR3	90
`reorSetChi::[Int]->[OB]->[OB]` set->chi->reoriented chi at set	86
`reorSetR3::[Int]->[(Tu,OM2)]->[(Tu,OM2)]` reorientation of uniform hyperline sequences in rank 3	90
`res::Int->[[Int]]` function for Petersen graph investigation	64
`res1::[Int]` matroid polytope corresponding to a Shemer sphere	207
`rest::[[([Int],OM2)]]` resulting hyperline sequence, rank 3, 5 elements	88
`resultCocircuits::[[Char]]` string of characters for cocircuit result	64
`si::Int->Int->[([Int],Int)]->Int` used in gpCheckR2, consR5, contra	116, 148
`sideOfHplane::Int->[Int]->Or` used in chi2Coc	126
`signH::Int->[(Tu,Or)]->Int`	173
`signRevR3::[(Tu,OM2)]->[(Tu,OM2)]` for sign reversal in rank 3	90
`signRevChi::[OB]->[OB]` chi->chirotope with signs of chi reversed	85
`signRevRow::(Tu,OM2)->(Tu,OM2)` for sign reversal in rank 3	90
`signsForExt::Int->Int->[Or]->[Or]`	176
`signsHead::[(Tu,Or)]->[[Int]]->[[Int]]` inserting the sign in the head of sublist	79, 173
`signT::OF->Int->Int` used in function signsTail	78, 173
`signsTail::OF->[[Int]]->[[Int]]` ordering function->list->inserted sign in tail of list	78
`signsTail::[(Tu,Or)]->[[Int]]->[[Int]]`	173

Table B.2   *(cont.)*

Function of Type Short explanation	Page
signVector::[OE]->GR->[Char] oriented edges->graph->list of signs of edges	61
siIn::[[Int]]->[(Tu,Or)]->[[Int]] complete sign insertion	79, 173
siIn::[OB]->[[Int]] prep. 2 of h.d.->chi->prep. 3 of h.d.	79
simplicialFacetsChi::[OB]->[[Int]] chi->facets of uniform chirotope	118
sphereA425::[[Int]] facets of an Altshuler 3-sphere	190
sts::[(Int,[[Int]])] start stars for K12 embeddings on a genus 6 surface	250
subfacets::[[Int]]->[[Int]] simplicial facets of 3-sphere->subfacets	147
subMA::[Int]->MA->MA row indices->matrix->submatrix	119
symmetry::Int->Int	225
t2gr::[Int]->GR->GR function for Petersen graph investigation	64
tailTup::Int->[[Int]]	176
triangle2Edges::[Int]->[[Int]]	233
ts::[[Int]] function for Petersen graph investigation	64
tu2q::Int->Int->[Int]->Int used in function dC for finding a dual chirotope	43
tup::Int->[Int]->[[Int]] Haskell primer function	296
tuples::Int->Int->[[Int]] r->n->all r-tuples of [1..n]	14, 168
tuplesL::Int->[Int]->[[Int]] r->list->all r-tuples of list	14, 168
twoEdges::[[Int]]->Bool	233
twoVertices::[Int]->Bool	233
vs2gr::[Int]->GR->GR vertices->graph->subgraph without vertices	62
vComp::[Int]->GR->[Int] vertices->graph->vertices of same component	62
xProd::MA->[Integer] exterior product in dimension 3	55

Appendix C   List of symbols

B^d	d-dimensional closed ball
$\mathbb{R}^d$, E^d	Euclidean d-space
$\langle .,. \rangle$	scalar product
$\|.\|$	euclidean norm
$|.|$	determinant
lin	linear hull
aff	affine hull
conv	convex hull
vert	set of vertices
$A + B$	Minkowski sum of A and B
int	interior
bd	boundary
relint	relative interior
S^{d-1}	$(d-1)$-sphere
$SO(n)$	special orthogonal group
$O(r)$	orthogonal group
$G(n, k)$	Grassmannian
$\mathcal{M}$	oriented matroid
$\chi(M)$	chirotope of matrix M
$(E, \chi)(M)$	chirotope of matrix M, E set of row indices of M
$\Lambda(n, r)$	r tuples out of n
$\lfloor . \rfloor$	floor function
$\binom{n}{r}$	binomial coefficient
$\mathcal{B}$	set of bases
$(E, \mathcal{B})$	matroid
$[i, j, k]$	bracket, determinant
C_m	cyclic group of order m
K_n	complete graph with n vertices
f_j	number of j-faces
$\{i, j\}_k$	regular map, i-gons, j meeting at a vertex, $k = $ Petrie polygon length

References

Altshuler, A., Bokowski, J., and Schuchert, P. (1994). Spatial polyhedra without diagonals, *Israel J. Math.*, **88**:373–96.

(1996). Neighborly 2-manifolds with 12 vertices, *J. Comb. Theory A*, **75**:148–62.

Altshuler, A., Bokowski J., and Steinberg, L. (1980). The classification of simplicial 3-spheres with nine vertices into polytopes and non-polytopes, *Discrete Math.*, **31**:115–24.

Bachem, A. (1983). Convexity and optimization in discrete structures, in *Convexity and Its Applications* (P. M. Gruber and J. M. Wills, eds.), Basel/Boston/Stuttgart, Birkhäuser Verlag.

Bachem, A. and Kern, W. (1992). *Linear Programming Duality: An Introduction to Oriented Matroids*, Berlin/Heidelberg, Springer-Verlag.

Barthel, G., Hirzebruch, F., and Höfer Th. (1987). *Geradenkonfigurationen und Algebraische Flächen*, Braunschweig/Wiesbaden, Vieweg.

Basu, S., Pollack, R., and Roy, M.-F. (2003). *Algorithms in Real Algebraic Geometry*, Springer-Verlag.

Biermann, M. (1997). Computerunterstützte Erzeugung orientierter Matroide im Rang 4, Diplom thesis, Darmstadt University of Technology.

Björner, A., Las Vergnas, M., Sturmfels, B., White, N., and Ziegler, G. M. (1993). *Oriented Matroids. Volume 46 of Encyclopedia Math. Appl.*, Cambridge, Cambridge University Press. Second edn 1999.

Blackburn, J. E., Crapo, H. H., and Higgs, D. A. (1969). A catalogue of combinatorial geometries. University of Waterloo, Waterloo, Ontario.

Bland, R. G. (1974). Complementary orthogonal subspaces of R^n and orientability of matroids, Ph.D. thesis, Cornell University.

Bland, R. G. and Las Vergnas, M. (1978). Orientability of matroids, *J. Comb. Theory, Ser. B.*, **24**:94–123.

Bohne, J. (1992). Eine kombinatorische Analyse zonotopaler Raumaufteilungen, Ph.D. dissertation, University Bielefeld.

Bokowski, J. (1987). Aspects of computational synthetic geometry, II: Combinatorial complexes and their geometric realization – an algorithmic approach. *Proc. of Computer-aided Geometric Reasoning*, INRIA, Antibes (France).

(1989). A geometric realization without self-intersections does exist for Dyck's regular map, *Discrete Comput. Geom.*, **4**:583–9.

(1991). On the geometric flat embedding of abstract complexes with symmetries, Symmetry of discrete mathematical structures and their symmetry groups (K. H. Hofmann and R. Wille, eds.), *Research and Exposition in Mathematics*, **15**:1–48, Berlin, Heldermann.

(1993). oriented matroids, in *Handbook of Convex Geometry* (P. Gruber and J. M. Wills, eds.), North-Holland, 555–602.

(1994). On recent progress in computational synthetic geometry, in *POLYTOPES: Abstract, Convex and Computational* (T. Bisztriczky, P. McMullen, R. Schneider, and A. Ivić Weiss, eds.), C: Mathematical and Physical Sciences, **440**:335–58, Kluwer Academic Publishers.

(2001). Effective methods in computational synthetic geometry, in *Automated Deduction in Geometry, Springer LNAI 2061*, (J. Richter-Gebert and D. Wang, eds.).

Bokowski, J. and Brehm, U. (1985). A new polyhedron of genus 3 with, 10 vertices. Colloquia Mathematica Societas János Bolyai **48**. Intuitive Geometry, Siófok, 105–16.

(1989). A polyhedron of genus 4 with minimal number of vertices and maximal symmetry, *Geometriae Dedicata*, **29**:53–64.

Bokowski, J., Cara, P., and Mock, S. (1999). On a self dual 3-sphere of Peter McMullen, *Periodica Mathematica Hungarica*, **39**:17–32.

Bokowski, J. and Eggert, A. (1991). All realizations of Möbius' torus with 7 vertices, *Topologie Structurale*, **17**:59–78.

Bokowski, J., Ewald, G., and Kleinschmidt, P. (1984). On combinatorial and affine automorphisms of polytopes, *Israel J. Math.*, **47**:123–30.

Bokowski, J. and Garms, K. (1987). Altshuler's sphere M_{425}^{10} is not polytopal, *European J. Comb.*, **8**:227–9.

Bokowski, J. and Guedes de Oliveira, A. (1990). Simplicial convex polytopes do not have the isotopy property, *Portugaliae Mathematica*, **47**:309–18.

(2000). On the generation of oriented matroids, *Discrete Comput. Geom.*, **24**:197–208.

Bokowski, J., Guedes de Oliveira, A., and Richter-Gebert, J. (1991). Algebraic varieties characterizing matroids and oriented matroids, *Advances in Math.*, **87**:160–85.

Bokowski, J., Guedes de Oliveira, A., Thiemann, U., and Veloso da Costa, A. (1996). On the cube problem of Las Vergnas, *Geometriae Dedicata*, **63**:25–43.

Bokowski, J., King, S., Mock, S., and Streinu, I. (2005). The topological representation of oriented matroids, *Discrete Comput. Geom.*, **33**(4):645–68.

Bokowski, J. and Kollewe, W. (1992). On representing contexts in line arrangements, *Order*, **8**:393–403.

Bokowski, J., Laffaille, G., and Richter-Gebert, J. (1989). 10 point oriented matroids and projective incidence theorems, unpublished.

Bokowski, J., Mock, S., and Streinu, I. (2001) On the Folkman–Lawrence topological representation theorem for oriented matroids of rank 3, *European J. Comb.*, **22**:601–15.

Bokowski, J. and Pisanski, T. (2002). Oriented matroids and complete graph embeddings on surfaces, *J. Comb. Theory, Ser. A.*, submitted.

Bokowski, J. and Richter-Gebert, J. (1990). On the finding of final polynomials, *European J. Comb.*, **11**:21–34.

Bokowski, J., Richer-Gebert, J., and Sturmfels, B. (1990). Nonrealizability proofs in computational geometry, *Discrete Comput. Geom.* **5**:333–50.

Bokowski, J. and Rohlfs, H. (2001). On a mutation problem for oriented matroids, *European J. Comb.*, **22**:617–26.

Bokowski, J. and Schewe, L. (2002). On Szillasi's torus, Symmetry: Culture and Science, **13**(3–4), 231–40.

(2005). *There are no realizable* 15_4 *and* 16_4 *configurations*, 60th birthday volume of Tudor Zamfiresw, 10 pp. (to appear)

Bokowski, J., Roudneff, J.-P., and Strempel, T.-K. (1997). Cell decompositions of the projective plane with Petrie polygons of constant length, *Discrete Comput. Geom.*, **17**:377–92.

Bokowski, J. and Schuchert, P. (1995a). Equifacetted 3-spheres as topes of nonpolytopal matroid polytopes, *Discrete Comput. Geom.*, **13**:347–61.

(1995b). Altshuler's sphere M^9_{963} revisited, *Siam J. Disc. Math.*, **8**:670–7.

Bokowski, J. and Shemer, I. (1987). Neighborly 6-polytopes with 10 vertices, *Israel J. Math.*, **58**:103–24.

Bokowski, J. and Sturmfels, B. (1986). On the coordinatization of oriented matroids, *Discrete Comput. Geom.*, **1**:293–306.

(1987). Polytopal and nonpolytopal spheres, an algorithmic approach, *Israel J. Math.*, **57**:257–71.

(1989a). An infinite family of minor-minimal nonrealizable 3-chirotopes, *Math. Zeitschrift*, **200**:583–9.

(1989b) *Computational Synthetic Geometry, Lecture Notes in Mathematics*, vol. 1355, Berlin/Heidelberg, Springer-Verlag.

Bokowski, J. and Wills, J. M. (1988). Regular polyhedra with hidden symmetries, *Mathematical Intelligencer*, **10**:27–32.

Brehm, U. (1983). A non-polyhedral triangulated Möbius Strip, *Proceedings of the American Math. Soc.*, **89**:519–22.

(1987). Maximally symmetric polyhedral realizations of Dyck's regular map, *Mathematika*, **34**:229–36.

Brown, M. (1962). Locally flat imbeddings of topological manifolds. *Ann. of Math. (2)*, **77**:331–41.

Cervone, D. (1994). Vertex-minimal simplicial immersions of the Klein bottle in three-space, *Geometriae Dedicata*, **50**:117–41.

(1996). Tight immersions of simplicial surfaces into three-space, *Topology*, **35**:863–73.

(1997). Tightness for smooth and polyhedral immersions of the real projective plane with one handle, in *Tight and Taut Submanifolds*, Proceedings of the Mathematics Sciences Research Institute (T. E. Cecil and S.-S. Chern eds.), 119–33.

(2000). A tight polyhedral immersion in three-space of the real projective plane with one handle, *Pacific Journal of Mathematics*, **196**:113–22.

(2001). A tight polyhedral immersion of the twisted surface of Euler characteristic-3, *Topology*, **40**:571–84.

Collins, G. (1975). Quantifier elimination for real closed fields by cylindrical algebraic decomposition, in *Automata Theory and Formal Languages* (H. Braghage ed.), vol 33 of Lecture Notes in Computer Science, Heidelberg, Springer, 134–63.

Coxeter, H. S. M. (1973). *Regular Polytopes*, 3rd edn., New York, Dover.

Coxeter, H. S. M. and Moser, W. O. J. (1980). *Generators and Relations for Discrete Groups*. Berlin, Springer-Verlag.

Császár, A. (1949). A polyhedron without diagonals, *Acta Sci. Math. Szeged*, **13**:140–2.

Dreiding, A., Dress, A. W. M., and Haegi, H. (1982). Classification of mobile molecules by category theory, *Studies in Physical and Theoretical Chemistry*, **23**, 39–58.

Dress, A. (1986). Chirotopes and oriented matroids. *Bayreuther Mathematische Schriften*, **21**, 55p.

Edmonds, J. and Mandel, A. (1982). *Topology of oriented matroids*, abstract 758-05-9, *Notices American Math. Soc.* **25**, A-410, 1978; also Ph.D. thesis of A. Mandel, University of Waterloo.

Engel, C. (1995). Zur Anwendung orientierter Matroide auf ein kombinatorisches Problem, Diplom Thesis, Darmstadt University of Technology.

Erdős, P. and Szegeres, G. (1935). A combinatorial problem in geometry, *Comp. Math.* **2**:463–70.

(1960). On some extremum problems in elementary geometry, *Ann. University Sci. Budapest Eötvös*, Sect. Nat. **3**:53–62.

Fenton, W. E. (1982). Axiomatic of convexity theory, Ph.D. thesis, Purdue University 98 pages.

Finschi, L. (2001). A graph theoretical approach for reconstruction and generation of oriented matroids. Ph.D. dissertation, ETH Zürich.

Folkman, J. and Lawrence, J. (1978). *Oriented matroids, J. Comb. Theory, Ser. B*, **25**:199–236.

Fomenko, A. (1994). *Visual Geometry and Topology*, Berlin Heidelberg, Springer-Verlag.

Forge, D. and Ramirez Alfonsin, J. L. (1998). Straight line arrangements in the real projective plane, *Discrete Comput. Geom.*, **20**:155–61.

Fukuda, K. (1982). Oriented matroid programming, Ph.D. thesis, University of Waterloo, thesis advisor Edmonds, J.

Ganter, B. and Wille, R. (1999). *Formal Concept Analysis, Mathematical Foundations*, Springer-Verlag.

Gardner, M. (1978). The minimal art, *Scientific American* **11**.

Gonzalez-Sprinberg, G. and Laffaille, G. (1989). Sur les arrangements simples de huit droites dans $\mathbb{R}P^2$, *C.R. Acad. Sci. Paris* Ser. I, **309**:341–4.

Goodman, J. E. (1997). *Pseudoline arrangements*, in *Handbook in Discrete and Computational Gemetry*, (J. E. Goodman and J. O'Rourke, eds.), Boca Raton, Fl., CRC Press, 83–110.

Goodman, J. E. and Pollack, R. (1980). Proof of Grünbaum's conjecture on the stretchability of certain arrangements of pseudolines. *J. Comb. Theory, Ser. A*, **29**:385–90.

(1981). A combinatorial perspective on some problems in geometry, *Congressus Numerantium* **32**:383–94.

(1984). *Semispaces of configurations, cell complexes of arrangements, J. Comb. Theory, Series A*, **37**:257–93.

(1993). Allowable sequences and order types in discrete and computational geometry, in *New Trends in Discrete and Computational Geometry* (J. Pach, ed.), *Algorithms and Combinatorics* **10**:103–34 Berlin/Heidelberg, Springer-Verlag.

Goodman, J. E. and O'Rourke, J. (eds.) (1997). *Handbook of Combinatorial and Computational Geometry*, CRC Press.

Gruber, P. M. and Wills, J. M. (eds.) (1993). *Handbook of Convex Geometry*, vols. A and B, North Holland.

Grünbaum, B. and Rigby, J. F. (1990). The real configuration 21_4, *J. London Math. Soc.* **2**, 336–46.

Grünbaum, B. (1972). Arrangements and Spreads, *Regional Conf., Vol. 10, Amer. Math. Soc.*, Providence, RI.

(2002). Connected ($n4$) configurations exist for almost all n – an update. *Geombinatorics* **12**, 15–23.

(2003). *Convex Polytopes*, 2nd edn, prepared by V. Kaibel, V. Klee, and G. Ziegler, Heidelberg, Springer Verlag.

(2005). Configurations of points and lines, *Fields Inst. Commun.* **48**, 1–46.

Guedes de Oliveira, A. and Carvalho, P. (2002). Intersection and linking in oriented matroids, Preprint.

Gugisch, R. (2005). Konstruktion von Isomorphieklassen orientierte Matroide, *Bayreuther Mathematische Schriften, Heft* **72**. ISSN 0172-1062. See also http://www.mathe2.uni-bayreuth.de/ralfg/origen.php

Gutierrez Novoa, L. (1965). *On n-ordered sets and order completeness*, *Pacific J. Math.*, **15**:1337–45.

Hall, C. and O'Donnell, J. (2000). *Discrete Mathematics Using a Computer*. London, Springer-Verlag.

Handa, K. (1990). A characterization of oriented matroids in terms of topes, *European J. Comb.*, **11**:41–5.

Harborth, H. (1985). Some simple arrangements of pseudolines with a maximum number of triangles, in *Discrete Geometry and Convexity, Annals of the New York Academy of Sciences*, **440**:30–1.

Heffter, L. (1891). Über das Problem der Nachbargebiete. *Math. Ann.* **38**, 477–508.

Hudak, P., Peterson, J., and Fasel, J. (1997). A gentle introduction to Haskell, `http://www.haskell.org`

Jaritz, R. (1996) Oriented matroids in terms of order functions, *Beiträge zur Algebra and Geometry*, 1–12.

Kalai, G. (1988). Many triangulated spheres, *Discrete Comput. Geom.*, **3**:1–14.

Kalhoff, F. B. (2000) Oriented rank three matroids and projective planes. *European Journal of Combinatorics.* **21**:347–65.

King, S. (2001). Polytopality of triangulations, Ph.D. dissertation, Université Louis Pasteur Strasbourg. `http://www-irma.u-strasbg.fr/irma/publications/2001/01017.shtml`.

Klein, O. (2002). Enumeration nicht-notwendig uniformer orientierter Matroide und untere Schranken für die zirkuläre Flusszahl. Diplomthesis, U. Köln.

Klin, M. H., Tratch, S. S. and Treskov, V. E. (1989). A graph-theoretic interpretation of 2D-configurations, in Proc. 4th All-Union Conf. *Methods and Programmes of Solving Optimization Problems on Graphs and Networks*. Vol. 1, Novosibirsk. 87–9 (in Russian).

Knuth, D. E. (1992). *Axioms and Hulls*, New York, Springer-Verlag.

Laffaille, G. (1989). Tape with all oriented matroids with up to 10 vertices, private communication.

Las Vergnas, M. (1974). Matroïdes orientables, 76 pages, *Texte Justificatif de la Note C.R. Acad. Sci. Paris*, **280** (1975), Ser. A, 61–4.

   (1975). Matroïdes orientables, *C.R. Acad. Sci. Paris, Ser. A*, **280**:61–4.

   (1978). Extensions ponctuelles d'une géométrie combinatoire orientée. Colloques internationaux C.N.R.S., No. 260, Problèmes combinatoires et théorie des graphes (Orsay 1976). EA. du C.N.R.S., Paris.

   (1978). Bases in oriented matroids, *J. Comb. Theory, Ser. B*, **25**:283–9.

   (1980). Convexity in oriented matroids, *J. Comb. Theory, Ser. B*, **29**:231–43.

   (1981). Oriented matroids as signed geometries real in corank 2, *Colloquia Mathematica Societatis János Bolyai*, 37. Finite and infinite sets, Eger (Hungary), 555–565.

Lavrenchenko, S. A. (1987). Irreducible triangulations of the torus, *J. Sov. Mat.*, **51**(5): 2537–43, transl. from *Ukr. Geom. Sb.* **30**, 52–62.

Lawrence, J. (1982). *Oriented matroids and multiply ordered sets*, *Linear Algebra Appl.*, **48**:1–12.

Levi, F. (1926). Die Teilung der projektiven Ebene durch Gerade und Pseudogerade, *Ber. Math.-Phys.Kl.Sächs.Akad.Wiss.*, **78**:256–67.

   (1929). *Geometrische Konfigurationen*, Leipzig, Hirzel.

Ljubić, D., Roudneff, J.-P., and Sturmfels, B. (1989). Arrangements of lines and pseudolines without adjacent triangles, *J. Comb. Theory Ser. A.*, **50**:24–32.

Lutz, F. H. (1999). Triangulated manifolds with few vertices and vertex-transitive group actions, Ph.D. thesis, Technical University Berlin.

(2001). *Császár's Torus*, Electronic Geometry Model No. 2001.02.069.
`http://www.eg-models.de/models/Classical\_Models/200./\`
`_preview.htm`

Lynch, T. (1982). The geometric body in Dürer's engraving Melencholia I, *Journal of the Warburg and Courtauld Institutes*, **45**, 226–32.

Mani, P. (1971). Automorphismen von polyedrischen Graphen, *Math. Ann.*, **192**, 297–303.

Mani-Levitska, P. (1972). Spheres with few vertices, *J. Comb. Theory, Ser. A*, **13**:346–52.

——— (1993). Characterization of convex sets, in *Handbook of Convex Geometry* (P. Gruber and J. M. Wills, eds.), North-Holland, 19–41.

McMullen, P. (1968). On the combinatorial structure of convex polytopes. Ph.D. thesis, University of Birmingham.

——— (1994). Modern developments in regular polytopes, in *POLYTOPES: Abstract, Convex and Computational* (T. Bisztriczky, P. McMullen, R. Schneider and A. Ivić Weiss, eds.), C: Mathematical and Physical Sciences, vol. 440, Kluwer Academic Publishers, 97–124.

McMullen, P. and Schulte, E. (2002). *Abstract Regular Polytopes*. Cambridge University Press, series Encyclopedia of Mathematics and its Applications.

Mihalisin, J. and Williams, G. (2002). Nonconvex embeddings of the exceptional simplicial 3-spheres with 8 vertices. *J. Comb. Theory, Ser. A*, **98**(1):74–86.

Mnëv, N. E. (1988). The universality theorems on the classification problem of configuration varieties and convex polytope varieties, in *Topology and Geometry* (O.Ya.Viro, ed.), Rohlin Seminar, 527–44, Volume 1346 of Lecture Notes in Math., Berlin/Heidelberg, Springer-Verlag.

Möbius, A. F. (1886) *Gesammelte Werke, Band II* (F. Klein, ed.), Verlag M. Sändig, Wiesbaden, 1886; reprint: Verlag S. Hirzel, Stuttgart, 1967.

Negami, S. (2004). Triangulations, in *Handbook of Graph Theory*, (J. L. Gross and J. Yellen, eds.), 737–60.

Paffenholz, A. (2004). New polytopes derived from products. Manuscript.

Perez Fernandez da Silva, I. (1995). Axioms for maximal vectors of an oriented matroid: A combinatorial characterization of the regions determined by an arrangement of pseudohyperplanes. *European J. Comb.*, **16**, 125–45.

Peyton Jones, S., ed., (2003). *Haskell 98 Language and Libraries, The Revised Report*, Cambridge University Press.

Pfeifle, J. and Ziegler, G. M. (2004). Many triangulated 3-spheres. *Math. Annalen* **330**:829–37.

Reinhardt, C. (1885). Zu Möbius' Polyedertheorie. Berichte über die Verhandlungen der königlich sächsischen Gesellschaft der Wissenschaften zu Leipzip, mathematisch-physische Classe 37, 106–25.

Richter-Gebert, J. (1989). Kombinatorische Realisierbarkeitskriterien für orientierte Matroide. Diplom thesis, Darmstadt University of Technology, Mitt. Math. Sem. Gießen, **194**:1–112.

——— (1991). On the realizability problem of combinatorial geometries, Ph.D. thesis, Darmstadt University of Technology.

——— (1993). Oriented matroids with few mutations. *Discrete Comput. Geom.*, **10**:251–69.

——— (1996). *Realization Spaces of Polytopes*, Lecture Notes in Mathematics, vol. 1643, Berlin/Heidelberg, Springer-Verlag.

Richter-Gebert, J. and Kortenkamp, U. H. (1999). *Cinderella – The Interactive Geometry Software*, Software and 143 pp. manual. Heidelberg, Springer-Verlag, `http://www.cinderella.de`.

Richter-Gebert, J. and Ziegler, G. M. (1994). Zonotopal tilings and the Bohne-Dress theorem, in *Proc. Jerusalem Combinatorics '93* (H. Barcelo, G. Kalai, eds.), *Contemporary Math.*, **178**: 211–32, American Math. Soc.

(1995). Realization spaces of 4-polytopes are universal. *Bull. American Math. Soc.*, **32**:403–12.

(1997). *Oriented matroids*, in *Handbook in Discrete and Computational Gemetry*, (J. E. Goodman and J. O'Rourke, eds.), Boca Raton, Fl., CRC Press, 111–32.

Ringel, G. (1956). Teilungen der Ebene durch Geraden oder topologische Geraden, *Math. Zeitschr.*, **64**:79–102.

(1959). *Färbungsprobleme auf Flächen und Graphen*, VEB Deutscher Verlag der Wissenschaften, Berlin.

(1974). *Map Color Theorem*, Berlin New York, Springer.

Ringel, G. and Youngs, J. W. T. (1968). Solution of the Heawood map-coloring problem, *Proc. Nat. Acad. Sci. USA*, 60:438–45.

Robertson, N., Sanders, D., Seymour, P., and Thomas, R. (1997). The four color theorem. *J. Comb. Theory, Ser. B* **70**, 2–44, Article No. TB971750

Roudneff, J. P. (1986). On the number of triangles in simple arrangements of pseudolines in the real projective plane. *Discrete Math.*, **60**:234–51.

(1996). The maximum number of triangles in arrangements of pseudolines, *J. Comb. Theory, B*, **66**:44–74.

Roudneff, J. P. and Sturmfels, B. (1986). Simplicial cells in arrangements and mutations of oriented matroids, *Geometriae Dedicata*, **27**:153–70.

Santos, F. (2002). *Triangulations of Oriented Matroids*, Memoirs of the American Math. Soc., Vol. 156, No. 741. Oxford University Press. www.oup.com

Scholl, P., Schürmann, A., and Wills, J. M., (2002). Polyhedral models of Felix Klein's group, *Mathematical Intelligencer*, **24**,(3):37–42.

Schulte, E. and Wills, J. M. (1985). A polyhedral realization of Felix Klein's map $\{3, 7\}_8$ on a Riemann surface of genus 3, *J. London Math. Soc. (2)* **32**:539–47.

Shannon, R. W. (1979). Simplicial cells in arrangements of hyperplanes. *Geometriae Dedicata*, **8**:179–87.

Shemer, I. (1982) Neighborly polytopes, *Israel J. Math.* **43**:291–314.

Simutis, J. (1977). Geometric realizations of toroidal maps, Ph.D. dissertation, University of California at Davis.

Schneider, R. (1993). *Convex Bodies: The Brunn–Minkowski Theory, Encyclopedia of Mathematics, Vol. 44*, Cambridge, Cambridge University Press.

Schuchert, P. (1995). Matroid–polytope und Einbettungen kombinatorischer Mannigfaltigkeiten, Ph.D. thesis, Darmstadt University of Technology.

Schuster, P.-K. (1991). Melencholia I, Dürer's Denkbild, Gebrüder Mann, Berlin, 679 pages.

Steinitz, E. (1922). Polyeder und Raumeinteilungen, Encyclopädie der mathematischen Wissenschaften, Band 3 (Geometrie), Teil 3AB12, 1–139.

Steinitz, E. and Rademacher, H. (1934). *Vorlesungen über die Theorie der Polyeder*, Berlin, Springer, reprint Springer-Verlag 1976.

Sturmfels, B. (1987). Aspects of computational synthetic geometry, I: Algorithmic coordinatization of matroids, *Proceedings of Computer-aided Geometric Reasoning, INRIA, Antibes (France)*.

Swartz, E. (2002). Topological representations of matroids, Preprint Cornell University, 22 pages.

Szilassi, L. (1986). Regular toroids. *Structural Topology*, **13**:69–80.

Thompson, S. (1999). *The Craft of Functional Programming*, Addison-Wesley, second edition.

Vince, A. (2004). Maps, in *Handbook of Graph Theory* J. L. Gross and J. Yellen, eds., 696–721.

Welsh, D. J. A. (1976). *Matroid Theory*, London, Academic Press.

Wills, J. M. (1987). On regular polyhedra with hidden symmetries, *Results in Math.*, **12**:450–58.

Wilson, S. E. (1976). New techniques for the construction of regular maps. Ph.D. dissertation, University of Washington.

Ziegler, G. M. (1994). *Lectures on Polytopes, Graduate Texts in Mathematics*, New York, Springer-Verlag, revised edition 1998.

(1996). Oriented matroids today, *The Electronic Journal of Combinatorics 3*, DS No. 4.

Index

Printed in the United States
by Baker & Taylor Publisher Services